AGRI Food Crops

NIPA® GENX ELECTRONIC RESOURCES & SOLUTIONS P. LTD.
New Delhi-110 034

AGRI Food Crops

(Processing, Value Addition, Packaging & Storage)

by

Sasi Kumar, R

Assistant Professor
Deptt. of Food Science & Nutrition
Central Agricultural University
Tura, Meghalaya

Sivakumar, P.S.

Scientist (SS)
Regional Tuber Crops Research Institute
Bhubaneshwar, Odisha

NIPA® GENX ELECTRONIC RESOURCES & SOLUTIONS P. LTD.

New Delhi-110 034

NIPA® GENX ELECTRONIC RESOURCES & SOLUTIONS P. LTD.

101,103, Vikas Surya Plaza, CU Block
L.S.C. Market, Pitam Pura, New Delhi-110 034
Ph : +91 11 27341616, 27341717, 27341718
E-mail: newindiapublishingagency@gmail.com
www: www.nipabooks.com

For customer assistance, please contact
Phone: + 91-11-27 34 17 17
Fax: + 91-11- 27 34 16 16
E-Mail: feedbacks@nipabooks.com

ISBN: 978-81-96079-01-7

Composed and Designed by NIPA®.

Visit us at: ouat.ac.in
Tel : (0674)2397700 (O), 2561606 (R)
FAX : 067-2397780
Email : ouat_dproy@yahoo.co.in/ouat_dproy@rediffmail.com
ORISSA UNIVERSITY OF AGRICULTURE & TECHNOLOGY
BHUBANESWAR-751003, ODISHA

Prof. D.P. Ray, Ph.D
VICE CHANCELLOR

FOREWORD

India is the second largest producer of food in the World next to China, which caters to the food needs of over 120 crore people everyday. The food processing industry is one of the largest industries in India ranking fifth in terms of production, consumption and export. Despite being the largest food producer, only 6% of the perishable commodities produced in our country are converted into value added products. The Ministry of Food Processing, Govt., of India envisaged an increase in processing level of perishable commodities from 6% to 12%, value addition from 20% to 35% and increase in India's share in global food from 1.5% to 3% by 2015 and is directing its efforts to achieve these levels.

With the rapid increase in per capita income and purchasing power along with increased urbanization, improved of living, the consumers are able to spend substantial amount on processed foods. Currently, it is estimated that over 300 million upper and middle class people regularly consume the processed food. The growing number of nuclear families and increasing levels of double income families have created a need for convenience foods too especially the urban Indians. Realizing this trend, several Indian and Multi-national companies offer a variety of ready-to-cook and ready-to-eat products which can save time and effort spent in cooking. Indigenous foods like *idli, dosa, roti, upma, halwa* etc. are also made into ready-to-eat from which slowly catching up the urban food market.

Diet-related, non-communicable disease like obesity, coronary heart disease are also in the rise due to sedentary life styles especially among urban Indians. This situation created a sense of vulnerability among urban Indians and most of them are now becoming health consicous and started consuming healthy functional foods. The Indian health food market is growing at 25-30% annually and expected to double in the few years.

Despite achieving huge strides in the convenience and health food development that resulted in an array of innovative products, there are very few books in the market that market that can provide comprehensive informtion one these specialized food categories. The book written by Mr. R. Sasi Kumar and Dr. P. Sethuraman Sivakumar entitled "Agri Food Crops " provides a comprehensive account on the preparation and production convenience of healthy food products. Besides, the book elaborates on the post-production aspects like sensory evaluation and marketing that are essential for success of these foods in the commpetitive Indian market. A chapter on preparing projects for bankable food products can help the prospective entrepreneurs to seek funding from credit Institutions to establish food industries in their area.

I congratulate both the authors for their commendable work by compling valuable information into a readable text book. I hope this book will serve as a valuable text book for students, teachers and scientists of Food Technology, Postharvest Technology, Food Processing & Preservation, Home Science and allied courses at the Graduate and Post Graduate levels.

28/12/11

(D.P. Ray)

PREFACE

The book is written to meet the needs of students of Indian Agricultural Universities pursuing courses in Food Technology, Postharvest Technology, Food Processing & Preservation, Home Science and allied courses, at the graduate and post graduate levels. The book currently used as text and reference books, mostly written by western authors, do not satisfy the needs of Indian students, which are important in ours. Our book eliminates this imbalance

The production, storage, processing and utilization of agro based food crops are discussed in details from the Indian context. Convenience and Health food products from cereals and millets, pulses and oilseeds, fruits and vegetables and milk based food products like idli, dosa, chapti, fruit juice, ready soup mix and milk based products have received special attention. Convenience and health foods from different source, which are important to overcome malnutrition of infant and young children of developing countries like India, are considered. Indian food laws and the role of Indian Standards Institute in regulation of food standards are discussed while production of convenience and health foods.

The book gives a comprehensive account of preparation and production convenience and health food products. It consists of 12 chapters, deal with available of local Agro based food crops resource, for producing and developing into various convenience food products, includes standard procedure, formulation and development of value added products from different combination. Chapter 12 deals Project preparation, how the local people can start food related project using local available resource, includes minimum financial requirements, fund allocation into different heads and judging a feasible project.

We trust that the book would be of interest to all students and scientists working in the field of Food Industry, Food processing and Preservation related research institutes. Suggestions for improvements of the book are welcome.

The authors wish to convey their grateful thanks to Dr.P.Banumathi, Dean, Home Science College & Research Institute, Tamil Nadu Agricultural University, Madurai, Tamil Nadu, India for writing the Foreword. Our special thanks to Dr.Kanchana, Associate Professor, Home Science College and Research Institute, Tamil Nadu Agricultural University, Madurai, Dr.J.T.Sheriff, Principle Scientists (Process Engineering), Central Tuber Crop Research Institute, ICAR, Thiriuvanathapuram, Kerala, India and Prof.Basanti Baroova, Head, Department of Food and Nutrition, Assam Agricultural University, Jorhat, Assam, for help at various stages of the preparation of the manuscript.

SASI KUMAR,R

SIVAKUMAR, P.S.

CONTENTS

Foreword *v*

Preface *vii*

List of Figures *xiii*

List of Tables *xv*

1. CEREALS, MILLETS AND PULSES BASED FOODS....................... 1-56

- Importance of Foods 1
- Foods: Emerging Scenario 2
- Breakfast Cereals 3
- Shelf Stable Fried Products 3
- Extruded Foods 7
- Value Added Products from Millets 8
- Extrusion Technology 17
- Development Process of Extruded Products 19
- Development of Weaning Food/Supplemented Foods 23
- New Product Development (Dough and Batter) 28
- Wheat Based Foods 31
- Fermented Products 36
- Soy Bean Based Value Added Products 49
- Quality Evaluation of Pulses Based Instant Mixes 54
- *References* 55

2. **TRADITIONAL FOODS.. 57 - 75**

- Importance of Traditional Foods 57
- Traditional Foods 57
- Preparation Ready Mixes (Masala and Soup) 60
- Preparation of Instant Soup Mixes 63
- Convenience Sweat Ready Mix 67
- Quality Evaluation of Traditional Sweet Mixes 73
- *References* 75

3. **FRUITS AND VEGETABLES BASED FOODS................................ 77 -106**

- Importance of Fruits and Vegetables in Human Life 77
- Value Added Products from Fruits 77
- Beverage Processing 99
- Fermentation 99
- Dehydration Process 103
- Pickle Production 104
- *References* 104

4. **TROPICAL ROOTS AND TUBER CROPS BASED FOODS......... 107-142**

- Introduction 107
- Cassava Flour And Starch 109
- Extraction of Starch From Dried Cassava Root 115
- Modified Starch 119
- Some Products from Modified Cassava Starches 120
- Sweet Potato Processing and Value Addition 128
- *References* 142

5. **MINOR FOREST PRODUCTS BASED FOODS........................... 143 -168**

- Tamarind 143
- Introduction 143
- Medicinal Uses of Tamarind 147
- Bael Fruits 147
- Processing of Value Addition 150
- Stevia 153

- Honey 154
- Honey Products 154
- Honey Processing 154
- Honey Production Process 157
- Honey Quality Control and Assurance 159
- Jackfruit 163
- Introduction 163
- *References* 167

6. MILK AND MILK PRODUCT BASED FOODS.......................... 169- 179

- Importance 169
- Traditional Indian Dairy Products 171
- Acid Precipitation Products 175
- *References* 179

7. FOOD ADDITIVES USED FOR FOODS 181- 188

- Importance of Food Additives 181
- Classification of Food Additives 181
- Coding of Food Additives 182
- Sensitivity of Food Additives 186
- Consumer Concerns and Issues of Food Additives186
- Food Additives: Approval Process 187
- *References* 187

8. FOOD PACKAGE DEVELOPMENT 189- 202

- Definition189
- Objectives of Packaging 189
- Method of Packaging 191
- Type of Food Packaging 192
- Package Material Suitable for Foods 192
- Package for Ready to Eat/ Ready to Cook Food 194
- High Moisture Food 198
- *References* 202

9. FOOD STORAGE AND QUALITY CONTROL.......................... 203 -211

- Importance of Food Storage 203

- Storage Methods 204
- Commercial Food Storage 206
- Emergency Preparation 206
- The Measures to Prevent Insect Infestation 207
- Quality Evaluation of Foods 207
- Microbial Load of The Foods 210
- *References* 211

10. FOOD LAWS AND REGULATION 213- 220

- Purpose of Food Law 213
- General Principles of Food Safety Risk Management 213
- Challenges of Food Regulation Law in Indian Food Industry 214
- Limitation of Current Food Laws 214
- Integrated Food Law and National Food Laws 215
- The New Food Bill- August 2006 217
- Food Safety 217
- National Food Security To Global Food Market 218
- *References* 220

11. PROJECT PREPARATION 221- 278

- Introduction of Food Industry 221
- Scope of The Projects 224
- Project on Manufacturing of Breakfast Cereal Food 225
- Project on Manufacturing of Protein Rich Biscuits 231
- Project on Manufacturing of Aseptic Juice Concentrate 234
- Project on Manufacturing of Amla Products 242
- Project on Manufacturing of Carbonated Fruit Beverages 245
- Project on Manufacturing of Dehydrated Vegetables 251
- Project on Manufacturing of Pasta Products 255
- Project on Manufacturing of Fruit Jam, Jelly and Marmalade 261
- Steps for Commercializing Value Added Food Business 265
- *References* 278

LIST OF FIGURES

FIGURE

Figure 1: Ragi vermicelli, 11

Figure 2: Ragi halwa, 13

Figure 3a: Flow diagram of Ready to Use ragi beverage mix, 13

Figure 3b: Flow diagram of Ready to Use ragi beverage mix, 14

Figure 4: Flow diagram of ragi Nutrimix, 16

Figure 5: Flow diagram of extruded products, 21

Figure 6: Single screw extruder, 21

Figure 7: Process flow diagram of weaning food preparation, 25

Figure 8: Flow diagram for the preparation of Instant idli mix, 38

Figure 9: Flow diagram for preparation of instant adai mix, 43

Figure 10: Flow chart for the preparation of instant vada mix, 46

Figure 11: Flow chart for the preparation of instant bajji mix, 47

Figure 12: Flow diagram of Processing of papad 59

Figure 13: Flow chart for the preparation of Kheer mix, 68

Figure 14: Flow chart for the preparation of carrot halwa mix, 70

Figure 15: Flow chart for cake ready mix, 72

Figure 16: Flow chart for the preparation of ice cream, 75

Figure 17: Amla processing machineries, 83

Figure 18: Flow diagram of Beverages classification, 84

Figure 19: Flow diagram for preparation of squash, 87

Figure 20: Flow diagram for processing of RTS beverages, 88

Figure 21: Flow diagram for processing of cordial, 89

Figure 22: Flow diagram for processing of Wine, 91

Figure 23: Flow diagram for processing Jam, 92

Figure 24: Flow diagram for processing of Jelly, 94

Figure 25: Flow diagram for processing of Marmalade, 95

Figure 26: Diagram of Traditional still, 97

Figure 27: Diagram of Small still, 98

Figure 28: Diagram for Liquid filler, 102

Figure 29: Flow diagram for Drying/dehydration fruits/vegetables, 105

Figure 30: Diagram for Medium scale cassava processing unit, 114

Figure 31: Diagram for Modern cassava processing plant, 116

Figure 32: Flow diagram for traditional processing technique, 127

Figure 33: Process comparison between Improved and Traditional Processing Stages of Sweet Potato flour, 128

Figure 34: Value added products from Sweet potato, 129

Figure 35: Tamarind pulp, 143

Figure 36: Flow diagram for Tamarind rice ready, 145

Figure 37: Whole bael fruit, 149

Figure 38: Broken bael fruit, 149

Figure 39: Flow diagram for pre-processing of bael fruit, 151

Figure 40: Fresh Jackfruit pulp, 164

Figure 41: Jackfruit processing unit, 164

Figure 42: Whole and cut jackfruit, 165

Figure 43: Jackfruit in Jar, 165

Figure 44: Jackfruit chips package unit, 165

Figure 45: Flow diagram for Dries jackfruit flakes, 166

Figure 46: Flow diagram for Jackfruit Leather, 167

Figure 47: Flow chart of conversion of milk into traditional Indian dairy products,170

Figure 48: Layout plan for food processing unit, 268

Figure 49: Food Analysis lab layout plan, 268

Figure 50: Food Microbiology lab layout plan, 269

LIST OF TABLES

TABLE

Table 1: Nutrition composition of ragi, 9

Table 2: Ingredients of ragi noodles, 10

Table 3: Ingredients of ragi vermicelli, 11

Table 4: Ingredients *of* ragi Idiyappam, 11

Table 5: Ingredients of ragi paniyaram, 12

Table 6: Ingredients of ragi Halwa, 12

Table 7: List of machineries required for millets milling unit, 17

Table 8 : Nutrition details of weaning food/100 gram, 24

Table 9: Production stages of rice noodles, 29

Table 10: Production stages of tortilla, 30

Table 11: Processing stages of baked products, 33

Table 12: Equipments required for baked products, 34

Table 13: Oven temperatures for different baking products, 35

Table 14: Ingredients Combinations of rava dosa, 40

Table 15: Materials required for rice flour, 41

Table 16: Raw ingredients used for Instant adai mixes, 42

Table 17: Ingredients used for instant bajji mix, 45

Table 18: Ingredients used for Instant pakoda mix, 48

Table 19: Ingredients used for instant murukku mix, 49

Table 20: Preparation nectar from different fruits, 90

Table 21: Type of alcoholic beverages, 97

Table: 22: Process and stages of beverage preparation, 98

Table 23: Equipment required for beverage processing, 98

Table 24: Different type of storage requirements, 101

Table 25: Type of package in different containers, 101

Table 26: Nutrient Value of Sweet Potatoes as Compared with other crops, 126

Table 27: Potential of processed products from tamarind, 146

Table 28: List of machinery and equipment for tamarind processing unit, 146

Table 29: Nutrition Value Per 100 g of Edible Portion, 148

Table 30: Nutritive value comparison with amla fruit, 149

Table 31: List of machinery and equipment used for honey processing unit, 157

Table 32: Honey standards, 161

Table 33: Product lists for milk and milk products based convenience foods, 169

Table 34: Nutritional comparison between cow and buffalo milk, 171

Table 35: Composition of Lassi, 171

Table 36: Composition of Rabri, 172

Table 37: Composition of khoa made from cow and buffalo milk, 172

Table 38: Composition of Pannier made from cow and buffalo milk, 175

Table 39: Composition of Chhana made from cow and buffalo milk, 176

Table 40: Classes and function of food additives, 182

Table 41: Use of Various Packaging Laminates/Composites, 196

Table 42: Projections of Marketable surplus, 222

Table 43: Status of Processed fruits and vegetables Industry in India, 222

Table 44: Major thrust area for project preparation, 269

Chapter - 1

Cereals, Millets and Pulses Based Foods

Importance of Foods

The word "instant" has been used in widely different types of premixes to specify the products either as ready- to-eat food or semi processed foods. The instant premixes available in Indian market were classified as Indian products and Westernized or sophisticated products. The instant mixes included under Indian market products are spice mixes, snack mixes and desert and sweet meat mixes. The Westernized products (instant mixes) contained soup mixes, beverage mixes, desert mixes and soft drink mixes.

Convenience foods are gaining popularity in recent years as they are very easy to handle, require minimum storage space and are attractive. Convenience foods have emerged as a new set of products in the international market. Most of the traditional foods that are commercially available in our country are being prepared / processed by cottage industries and by some of the multinational companies. The various types of convenience foods/ instant mixes that are presently being processed, marketed and consumed all over the country. Supplementation with protein- rich sources and preparation of acceptable ready-to-eat snack foods would not only correct its nutritional inadequacies, but would also provide a variety.

Instant mixes were high in nutritive value especially needed for armed forces, light in weight, easy to carry and prepare. Rapid industrialization and urbanization and changes in eating habits of Indian people led to very high demand for ready-to- use snack foods. The convenience instant foods are meeting the urgent and exigency situations of offering hospitality to unexpected guests.

Changing life styles and values are at the root of considerable changes in eating habits over the last few years. Hand-in- hand with a strong demand for fast and take away foods, there has been a marked upturn in popularity for convenience foods in

India. The major reasons for this trend are: the decline of the family meal time, growth in disposal incomes (only in cases of city dwellers), the desire for more leisure time and demand for "foreign" or "sophisticated" dishes inspired by the media and increased travel.

In a busy and fast moving life where the housewife is also a bread- earner these instant products come in very handy, save time and take away the monotony of the kitchen. With the rapidly increasing standards of living these ready – to-use products are finding a place in our products line also, and in recent years, the demand for these is one on the increasing trend.

All places where supply of fresh food materials is logistically not feasible, troops are issued operational pack rations. These rations are mainly based on ready-to-eat or instant convenience food products packed in a manner that troops unaccustomed to cooked are able to prepare palatable and nutritious meals merely by warming or by mixing with hot water. These are particularly useful to troops who have to operate under inhospitable terrain and climate where even normal cooking facilities are not available.

In short, the needs for convenience foods are

1. Increased education and employment opportunities for women
2. Large number of employed couples
3. Increased industrialization and urbanization
4. Large floating population due to promotion of tourism
5. Better wages and consequently higher surplus incomes
6. Changing life styles and food habits of middle income groups
7. Better awareness about processed foods and above all better availability of convenience foods.

Foods- Emerging Scenario

The term "convenience foods" is used for a very heterogeneous group of foods which vary in composition, shape, size, method of preparation and processing and even with regard to their functions in the diet. These literally range from simple fried to roasted nuts to ready mixes to canned and frozen foods to sophisticated warm-and-serve type TV dinners. Convenience foods can be defined as those products in which all or a significant portion of their preparation has been transferred from the consumers kitchen to the processing plant. The major emphasis in this definition is on the quantum of convenience which should be significant and should provide adequate savings in time and labour in the consumers' kitchen.

All over the world during the last two decades, the convenience food market has witnessed breath-taking changes in quality and quantity of products available and the packaging and technology employed for their processing. In India too, many of the products which were hitherto marketed by food service establishments or small scale artisans are now marketed by multinational companies in attractive packages. Though the market has not grown too commensurate with the population, the changes are quite impressive keeping in view the low level of industrialization.

Breakfast Cereals

The cereal grains find an important use in the manufacture of breakfast cereals, most breakfast cereals are made from the endosperm of wheat, corn, rice, or oats. The endosperm may simply be broken or pressed, with or without toasting, to yield such uncooked cereals as farina and oatmeal.

But far more popular are the so-called ready to eat cereals. For these, the endosperm may be broken or ground into a mash, and then converted into flakes by squeezing the broken grits or mash between rollers. The mash also can be extruded into numerous shapes; or the endosperm may be kept intact as kernels to be puffed, as in the case of puffed rice. But in all cases, the flaked, formed, or puffed cereal must be oven-cooked and dried to develop toasted flavor and to obtain the crisp, brittle textures desired. This crispness requires that many ready-to-eat breakfast cereals be dried to about 2-5% moisture.

Shelf Stable Fried Products

Fried products form the largest group of convenience foods marketed in India. A large number of fried items are made from flours of wheat, bengalgram, blackgram and rice throughout the country. Frying not only aids in cooking, but it helps in the destruction of anti-nutritive factors, aids moisture evaporation and thereby imparts shelf- stability. Above all, it imparts crispness and fried aroma which make the fried products highly acceptable. The moisture contents in these products range from 2-6% and products equilibrate to 0.30aw. The fat contents vary from 25-37%. As a result of low moisture and very low water activity, the products do not support any microbial proliferation and remain stable for long periods provided moisture ingress is prevented by packing and storing in moisture proof packs. Because of very low moisture, the products are highly susceptible to peroxidation of fats which results in off flavours and rancidity in the product. Products fried in unsaturated vegetable oils become rancid faster than those fried in relatively saturated oils like palm oil and vanaspathi.

Both sweet and savoury fried products are much liked and prepared throughout the country. Despite their high popularity, their commercial marketing has not been taken place to any appreciable extent. Major constraint in their large scale production seems to be the non-availability of continuous deep fat frying equipments, and relatively high prices of quality vegetable oils which are necessary for producing quality products of reasonably long shelf life. It is also rather surprising that no effort has been made in trying these highly acceptable and palatable products in nutritional programmes for school going children because these are concentrated sources of calories and their protein levels can also be enhanced by incorporating defatted oilseed or legume flours besides being highly acceptable.

(a) Shakarparas and namkeenparas: These are sweet and salty products prepared from stiff dough containing wheat flour, vanaspathi, water and sugar or salt, respectively. The dough is flattened into thin sheets, cut into desired shapes and fried in oil at temperature ranging from 140-160°C. The moisture in the product varies from 2-5% while the fat content varies from 27-37%. Shakarparas and namkeenparas remain stable for one year when fried in vanaspati and packed in paper-Al-foil-polyethylene laminate packs.

(b) Fried products from bengal gram: These are salted and spiced crispies in varied shapes prepared from bengal gram flour dough by extruding through orifices and deep-fried in vegetable oils. Products termed Sev, Ganthiya, Papdi, Boondi etc. are highly relished as such, and are also used in the preparation of curries from curd. Due to high cost of Bengal gram flour, some attempts have also been made to replace part of bengal gram flour with defatted soyflour and maize flour with encouraging results. Horse gram flour along with bengal gram flour is used commercially for preparing Bhujiya. Despite good internal demand and also good potential for export, there have been practically no efforts in atomization of the process of extrusion and frying. Mechanized extrusion of Sevs (Bhujiya) seems to harden their texture and reduces the fat absorption during frying operation.

(c) Fried products from rice and legumes: Chakli, Murukku, Tengolal, Muchorai and Kodhale are some of the popular fried products prepared from rice and legume flours while Chakli and Tengolal are based on rice and black gram in the proportion of 4:1 and 3:1 respectively; Muchorai is prepared from rice and green gram dhal flours in nearly equal proportion and Kodbale is prepared from rich flour and roasted bengal gram dhal flour having salt, spices and optionally coconut powder. Stiff dough is extruded through a hand operated press and fried. Crispness and soft texture are the most desirable characteristics, and these are mostly determined by the dough composition.

(d) Fried dhals: Fried dhals and whole legumes are prepared by soaking in water along with salt and sodium bicarbonate solution, and frying in vegetable oils and mixing with spices. Bengal gram, green gram, horse gram and black gram dhals are most frequently used. Fried dhals are highly crisp. Storage studies conducted in different packaging materials revealed oxidative off flavours as the major cause of storage deterioration though increase in peroxide value did not correlate with acceptability of the stored product. Fried green gram dhal was found to remain stable for 120 days at 37°C and for 210 days at 27°C when packed in paper-Al-foil polyethylene laminate pouches. Despite fried dhal being highly popular as snack food, it is rather surprising that its production has remained in the hands of skilled artisans. Food processing industry has made little effort in mechanizing its production and bringing out innovative changes in its marketing and shelf stability.

Moist Fried Products

Products like Samosha, Cutlets, Vada, Pakora, Kachori, Bhaji etc., are consumed as snacks vary widely in India. These are prepared and marketed mostly by food service establishments or confectioners. Total sales of these products probably may surpass any other type of snack but because of their moisture content (15-40%), they are susceptible to microbial growth especially ropiness, fermentation and mould growth. These products continue to be prepared by labour intensive traditional methods and there has been practically to improvement in their method of manufacture. However, because of rapid urbanization, increasing sophistication and improved hygiene requirements, there is an urgent need for mechanized production of these products.

Popped or Puffed Cereals

A variety of popped or puffed cereals like popcorn, puffed rice (Kheel, khof, aralu, nelpuri), puffed sorghum (Kheel), puffed ragi and puffed barley (dhanj, satu) are marketed and consumed as snacks by all segments of population. While popcorn is popular among urban elite, other puffed cereals are more popular among middle and lower income segments of the population in regions where the specific cereal is predominantly grown. Also, while all popcorn is produced in electrically operated mechanized poppers and marketed in polyethylene packs, the other popped cereals are manufactured by small scale artisans by heating conditioned cereals in hot sand and marketed in loose by petty shopkeepers. All popped cereals are highly crisp, fragile and except ragi and barley, have fluffy white attractive appearance. But they are highly hygroscopic and tend to absorb moisture and become leathery, chewy and soggy, a greatest constraint in the traditional method of marketing. Blending with different flavours and marketing them in moisture-proof plastic film pouches provide enormous opportunities for increased acceptance and usage of puffed cereals.

Among other factors like colour, flavour, texture and other subjective characteristics, the expansion ratio i.e., the relative amount of kernel expansion on popping is given the largest weightage in selecting varieties for preparing puffed cereals. For efficient puffing, the temperature just before explosion must be sufficient to create a high enough water vapour pressure without burning the pericarp and the temperature increase must be fast enough to build up the required pressure before the water evaporates. Moisture content of the kernel has a pronounced effect on popping behaviour. The kernels which are too dry often pop up feebly.

In the preparation of puffed rice, paddy is equilibrated with moisture and then roasted in sand. The maximum expansion ratio has been observed to be only around 15-fold while optimum moisture varied between 13.5 –14.5% and optimum temperature was found to be 190-210°C for air puffing. Soaking paddy in salt solution improved expansion but parboiling as well as pressure parboiling considerably reduced puffing.

Though puffed ragi is not as widely used as puffed rice or popcorn, some work has been reported on the puffing quality of various varieties of ragi vis-à-vis their physico-chemical characteristics. Wide varietal variations in puffing quality of ragi have been observed but no consistent relationship could be established with grain amylose, protein content and bran thickness.

Despite wide range of puffed cereals in our country, data on their nutritional values and flavouring components are scanty. Also, no efforts have been made to mechanise their large scale production and market them in moisture-proof packaging. Major constraint seems to be the non-availability of low cost mechanized roaster, and filling and sealing machines. Developmental efforts in this direction as well as exploration of possibility of marketing puffed cereals with innovative flavours will go a long way in increasing their acceptance by a wider section of population.

Expanded Cereals

Expanded rice (murmura, puri, muri) is another traditional convenience food widely consumed in India either as such or with jaggery, roasted Bengal gram and shredded vegetables and spices. The product is mostly produced in home or cottage sector by skilled artisans. In the traditional process, the paddy is soaked in water preferably overnight until saturation, drained and then either steamed or dry roasted in sand for parboiling. The parboiled paddy is milled, salted and again roasted in sand for expansion. Recently, pressure parboiling method has also been tried for preparing expanded rice. In this process, paddy is soaked for short periods, drained and steamed under pressure to bring about gelatinization. The parboiled rice is then roasted in sand after salt and bisulphite treatment.

At present, dry heat parboiled rice is in vogue under commercial conditions. Switching over to pressure parboiling treatment will involve establishing additional boiler and retorting facilities involving considerable amount of additional expenditure. Also, pressure parboiling treatment (2.5 kg / cm^2 for 15 min.) results in amber coloured parboiled rice and consequently yellowish coloured of expanded rice. Among other factors, the moisture content of parboiled rice, roasting temperature, rice to sand ratio, and rice variety also influence the expansion ratio.

Beaten Rice

Flaked or beaten rice (Poha, Avalakki, Chivda) is also a very popular traditional product which is consumed either as snack after toasting or frying and spicing or after soaking in water and seasoning with spices and vegetables as an item of breakfast. Beaten rice is consumed relatively in larger quantities in Western and Southern parts of India. Thinner flakes having whitish colour are preferred as compared to shorter and thicker flakes with dark colour. Commercially beaten rice is prepared by soaking paddy in warm water (40-50°C) for 18-24 hrs. After draining off the soak water, the soaked paddy is roasted in sand (220-240°C) and subsequently flaked. Yield of the flaked rich is rather low in this process and ranges between 61-64%. Some attempts have been made to increase the yield by altering paddy curing and flaking conditions.

There is scope for reduction in breakage during transportation and marketing channels and colour and flavour of the product also need improvement. Also, there is need to develop value added ready-to-eat products from flaked rice which could be used either as a breakfast item or as a snack. A process for flaking of jowar (sorghum) by soaking in water, roasting in sand and light polishing in a polishing machine followed by flaking in a roller flaker has been developed at CFTRI, Mysore. Flaked jowar can be consumed as chivda after seasoning etc.

Extruded Foods

In India, many extruded foods are prepared either by housewives themselves or by skilled artisans from rice or rice and legumes and served as snack. These are mostly prepared by cooking rice flour with water and extruding the gelatinized paste through a hand operated press in the form of noodles, chords or ribbons and drying them in sun. Dried products remain stable for many months and just before serving, these are fried in oil resulting in many-fold expansion and a soft and crisp texture. Alternatively, extruded foods are also prepared by making a dough of rice and legume flour (black gram, green gram or Bengal gram flour) and extruding in hand operated press into different shapes and directly frying in vegetable oils.

Fried products as already stated remain stable for 1-2 months. Both the types of extruded products are generally prepared in homes but small quantities are also marketed by confectioners or in condiment stores. With the advent of mechanized cooker extruders, a few companies have started marketing extruded pellets (Fryums) which need frying before use and ready-to-eat expanded and spiced products in attractive packs. Both the types of products are receiving good consumer response in the market and a number of other companies are planning to bring out similar products.

(a) Extruded pellets: These are double extruded products prepared from wheat, rice or maize flours or their blends. After proper conditioning and equilibration, the ingredients are cooked in a primary extruder. The cooked material is then cooled, solidified and shaped on a secondary extruder. In the secondary stage, the temperature of the extruded cooking is kept below the boiling point of water of eliminate product expansion. Immediately after extrusion, the moisture content in pellets is around 20-25% which is brought to 7-8% by drying in a tunnel dryer. The dried pellets are packed in attractive moisture-proof packs of plastic films. Before consumption, these have to be fried in oil which imparts crispness and softness and about 4-10 times expansion in volume. One of the important advantages of pellet type products is that they have long shelf life and any off flavour generated during storage gets volatilized during frying operation.

Secondly, the product is not fragile and packaging requirements are relatively less stringent compared to ready-to-eat expanded products.

(b) Ready-to-eat expanded products: These products are manufactured using high pressure cooking and forming equipments. The products processed by this method, such as crax or corn curls etc., are merely toasted and dried and then coated with flavour either by dusting or by spraying with spice-oil mixture. Advantages with this type of product are that the level of the oil / fat in the product is kept very low and the products are ready-to-eat. The products can be marketed in different shapes and designs having different flavours.

Both pellet and expanded extruded products are produced from various cereal products (grits, flour, starches etc.,) or their blends. The extruded products are mainly depends on expansion, flavour, colour, moisture and temperature of extrusion. As is evident, rice and corn flour or meal is most expandable. While rice products have a bland flavour and can be given any desired additional flavour, the corn extruded products have a strong but well accepted flavour. Wheat and tapioca flours expand relatively less.

(c) Corn and tortilla chips: Corn and tortilla chips are popular as snacks in American continent. Both these products are prepared from corn which is cooked with dilute solution of lime by bringing to near boiling. The corn is then cooled and allowed by steeping under water for 8-16 hr. The cooking and steeping process loosens the outer hull and germ which are removed by washing. The washed kernels are ground between milling stones to produce a mass having 50-60% moisture.

To make corn chips the mass is extruded through a slit to give ribbon of dough which is cut off by a knife in rectangular chips. These fall directly in hot oil and deep-fried to a moisture content of around 2-4%. Tortilla chips are produced by rolling the mass into a thin sheet, cutting into desired shapes and finally the chips are transferred into open flame oven. The heat sets the structure, drives off the moisture and generates a mild caramelized flavour. The chips are then briefly fried to create a crisp texture, salted and packed. Being low in moisture, these are susceptible to rancidity development during long term storage. Though Indian public is not accustomed to these products, their technology of manufacture is simple and can be adopted even in cottage sector.

Value Added Products from Millets

The food processing sector is a highly fragmented industry, Broadly the food industry is classified into various sub-segments viz. : fruits and vegetables, milk and dairy products, juice and beverages, meat and poultry, marine products, grain processing, packaged or convenience food, packaged drinks and nutraceutical products.

A large number of entrepreneurs in food industry are small in terms of their production and operations, and are largely concentrated in the unorganized segment. Food processing units can be one of the priority sectors when it comes to GDP Issues.

R&D Supports for Food Processing Enterprises

Government of India has created CSIR, DRDO, and ICAR State Agriculture Universities across the nation with its own R&D facility to cater to the new food process techniques/technologies along with training of required manpower for the food processing industries. Government assistance The Government has introduced several schemes to provide financial assistance for setting up and modernizing of food processing units, creation of infrastructure, support for research and development and human resource development in addition to other promotional measures to encourage the growth of the processed food sector.

All the developed and fermented food mixtures when rehydrated were found to be organoleptically acceptable. The food has high starch and protein digestibility (in vitro) 78 to 96% and 47-55%, respectively, significant lowering of serum cholesterol and LDL – cholesterol level and reduced gastrointestinal side effects when used with ampicilline.

Value Added Health Food Products from Millets

Various value added products were developed and standardized using pearl millet, finger millet, barnyard millet and sorghum. Some health foods were also developed for diabetics utilizing pearl millet. The details of food products developed are summarized below:

Ragi or Finger Millets

Ragi is also known as finger millet. It constitutes a little over 25% of the food grains grown in India. Nutritionally it is almost as good as or better than wheat or rice. The major proteins of ragi are prolamins and glutenins and they appear to be adequate in all the essential amino acids. Ragi is rich in minerals especially calcium. It is also rich in fibre. It is also rich in phytate and tannin and hence interferes with mineral availability. It contains B-vitamins but is poor in B2. Malting of finger millet is a traditional process followed in India and is used in infant foods and in milk thickener formulations, conveniently called ragi malt.

The grain is also malted and the flour of the malted grain is used as nourishing good for infants and invalids. Malting releases the amylases which dextrinize the grain starch. An added advantage of malting ragi is in the production of an agreeable odour developed during the kilning of the germinated grain. Malted ragi flour is called 'ragi malt' and is used in the preparation of milk beverages. A fermented drink or beer is also prepared from the grain in some parts of the country.

Table 1: Nutrition composition of ragi

Food	Energy Kcal	Protein g	Fat g	Carbohy -drates g	Calcium mg	Iron mg	β-Caro-tene mcg	Thiamine mg	Riboflavin mg	Niacin mg
Ragi	328	7.3	1.3	72.0	344	3.9	42	0.42	0.19	1.1

The nutritive value of ragi is better than that of rice and other cereals. The husk forms 5.6 per cent of the weight of the grain. The average composition is also follows; moisture, 13.1; protein, 7.1; fat, 1.3; carbohydrates, 76.3; and mineral matter, 2.2 per cent. It is rich in calcium, phosphorus and iron, the calcium content is higher than in the common cereals and millets. The major proteins of ragi are prolamins and glutelins and they appear to be adequate in all the essential amino acids. Germinated finger millet is used to make weaning foods for infants.

Processing Steps

Milling

Ragi can be milled by wet conditioning. It can be steamed followed by milling in a hammer or plate mill or a roller flour mill.

Malting

Compared to other millets, ragi is most suitable from the stand point of product quality and enzyme release for malting. The malted ragi flour can be used along with germinated green gram flour to formulate a high calorie dense weaning food having excellent nutritional qualities. Ragi flour can be used with milk beverages. Parboiling of ragi helps in the quality of ragi dumpling by eliminating its slimy texture. Flour from puffed ragi has good flavour and can be used in snacks and supplementary foods in south India ragi is used as gruel, dumpling, roti, dosa or porridge.

Value Addition

Ragi Noodles

Extrusion cooking because of its low cost and continuous processing capability has been accepted as one of the most useful technologies during the recent years in the field of food processing.

Table 2: Ingredients of ragi noodles

Refined wheat flour	70 g
Ragi	30 g
Water	30 m l
Salt	2 g

Method

- Sieve refined wheat flour (control) and ragi flour blends in a BS 60 mesh sieve, steam for five minutes, cool and sieve again.
- Fill the flour in the mixing compartment of the pasta-making machine and blend with water and salt for 30 minutes and extrude.
- Steam the noodles for 5 minutes
- Allow to temper in room temperature for 8 hours
- Dry in a cabinet drier at 60 °C for 6 hours.

Ragi Vermicelli

Table 3: Ingredients of ragi vermicelli

Refined wheat flour	30 g
Whole wheat flour	40 g
Ragi	30 g
Water	30 ml
Salt	2 g

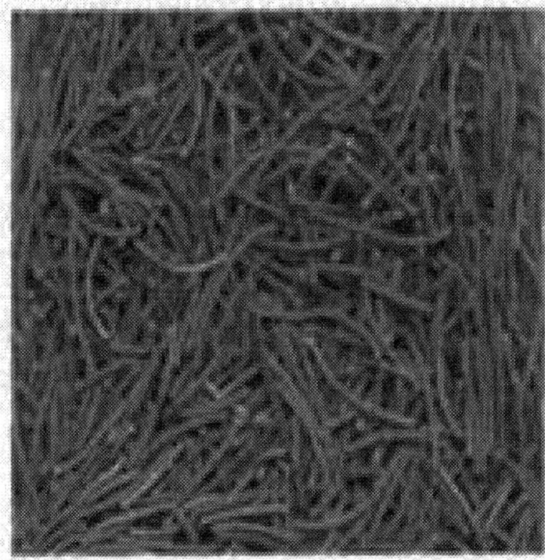

Figure 1: Ragi vermicelli

Method

1. Sieve refined wheat flour, whole wheat flour and ragi flour blends in a BS 60 mesh sieve, steam for five minutes, cool and sieve again.
2. Fill the flour in the mixing compartment of the pasta-making machine and blend with water and salt for 30 minutes and extrude.
3. Steam the vermicelli for 5 minutes
4. Allow to temper in room temperature for 8 hours
5. Dry in a cabinet drier at 60°C for 6 hours.

Ragi Idiyappam

Table 4: Ingredients of ragi Idiyappam

Rice flour	80 g
Ragi	30 g
Water	30 ml
Salt	2 g

Method

- Sieve rice flour and ragi flour blends in a BS 60 mesh sieve, steam for five minutes, cool and sieve again.
- Fill the flour in the mixing compartment of the pasta-making machine and blend with water and salt for 30 minutes and extrude.
- Steam the idiyappam for 5 minutes
- Allow to temper in room temperature for 8 hours
- Dry in a cabinet drier at 60°C for 6 hours.

Paniaram

Table 5: Ingredients of ragi paniyaram

Parboiled rice	50 g
Raw rice	50 g
Ragi flour	50 g
Black gram dhal	25 g
Fenugreek	2.5 g

Method

- The cleaned and milled rice, millets and other ingredients were washed with water, drained and soaked with fresh water for 6 hours.
- The soaked water was drained and the soaked ingredients were ground in an electrically operated stone grinder till the batter became smooth in texture.
- The batter was mixed with 2 per cent salt and left to ferment for 12 hours at room temperature.
- The paniyaram mould was greased with 2 tsp oil and the fermented batter was poured into the paniyaram moulds and cooked on each side till well cooked and golden brown. The paniyaram was cooled and evaluated.

Ragi Halwa

Table 6: Ingredients of ragi Halwa

Ragi flour	100 g
Powdered sugar	100 g
Ghee	100 g
Cardamom powder	1 pinch
Cashewnuts	10 g

Figure 2: Ragi Halwa

Method

- Fry ragi flour in half the amount of given ghee in a heavy bottom kadai
- Add coconut milk powder with the flour and cook the flour in water
- When it thickens add sugar
- When the halwa forms a mass add the remaining ghee
- Stir continuously till the halwa leaves the sides of the pan and the ghee separates from halwa
- Add the coarsely ground cashew nuts
- Spread on a greased tray and cut into pieces.

Pearl Millet

Minor millets are grown over seven million hectares of land in India, producing five million tons of grains. The richness of millet varieties in the dry lands of southern India is similar to the diversity seen in Africa. Finger millet alone accounts for 2.6 million hectares, producing 3 million tons and providing staple food for people in Karnataka, Tamil Nadu, Andhra Pradesh, Orissa, Maharashtra and Bihar. Millets in Indian diets are classified as coarse cereals with small grains having 2.1 – 7.1 gm/1000 grain weight. Well- filled grains have 1.4 – 5.1 ml/1000 grain volume. They have

Figure 3a) : Flow diagram of Ready to Use ragi beverage mix

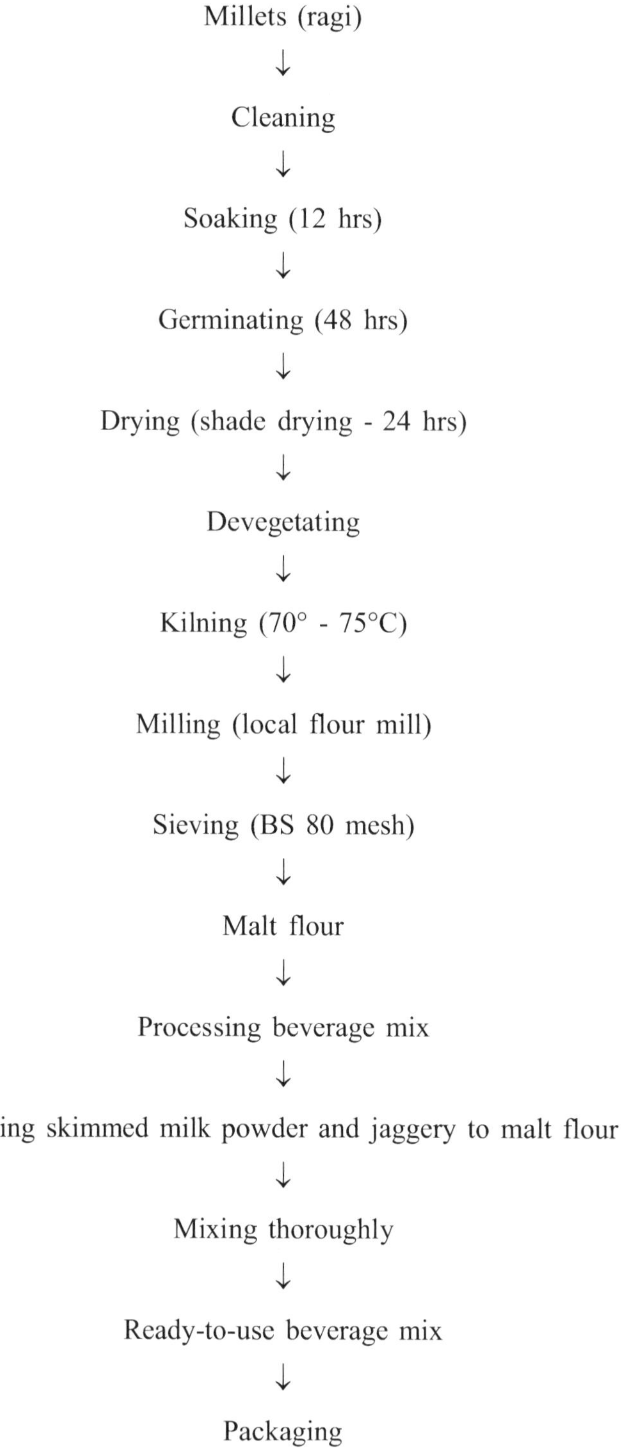

Figure 3b) : Flow diagram of Ready to Use ragi beverage mix

spherical to oval shape with colored seed coats. Millet is relished mostly by the rural population in India for its nutritional value, being a rich source of carbohydrates and minerals, such as calcium, phosphorous and iron. The major millet varieties in India are: (a) sorghum *(Sorghum bicolor)*; and (b) pearl millet *(Pennisetum typhoides)*.

Millets are rich source of nutrients. There are certain constraints like coarse skin, undesirable colour pigment presence of anti nutrients, poor keeping quality, etc. which hinder the utilisation of these millets. Various processing methods have been developed and standardised to overcome these constraints and improve the nutritive value of millets. These processing help to improve utilisation of millets for product development.

The processing also helps in increasing the shelf life of the processed pearl millets flour to 2-3 months from 5-8 days. The shelf life study indicates that the fat acidity, free fatty acids and lipase activity increased by almost four folds at the end of storage period in case of unprocessed pearl millet flour whereas the increase in these parameters was significantly lower in processed flour. During storage it was also noticed that acceptability of colour and appearance in raw pearl millet flour was much lower as compared to processed pearl millet flour.

1. Formulation of diabetic products/mixes – Pearl millet being a good source of dietary fibre and phyto-chemicals can be used for preparation of diabetic foods.

The process developed is a preparation of diabetic formulations with fenugreek and Bengal gram seed coat in pearl millet base. Before use the pearl millet flour is processed and fenugreeks were germinated and oven dried, all the components were ground and necessary components of sweetening and salt in case of salt biscuits were added to produce the various products given below.

- Preparation of *Chapatti* mix, *dhokla* mix, *idli* mix- *chapatti* mix and pasta were prepared using different proportions of bleached or unbleached pearl millet with Bengal gram seed coat.
- Pasta - prepared using different proportions of bleached or unbleached pearl millet with Bengal gram seed coat.

2. Development of supplementary foods: *Nankhatai, namkeen matar* and popped *ladoo*. Milk powder/soya bean/chickpea flour etc. was also added to improve its nutritive value of the supplements.
3. Development of baked products: Cake, biscuits, Soup sticks, Rusk, Salt biscuits. Biscuits were prepared using different proportions of bleached or unbleached pearl millet with Bengal gram seed coat.

All the products developed from various proportions of millets were found to be acceptable and nutritious. The products are not at all available in the market. Traditional products based on wheat flour/refined flour are available. Costs of pearl millet based products are almost 50% of traditional products and are superior in nutrients.

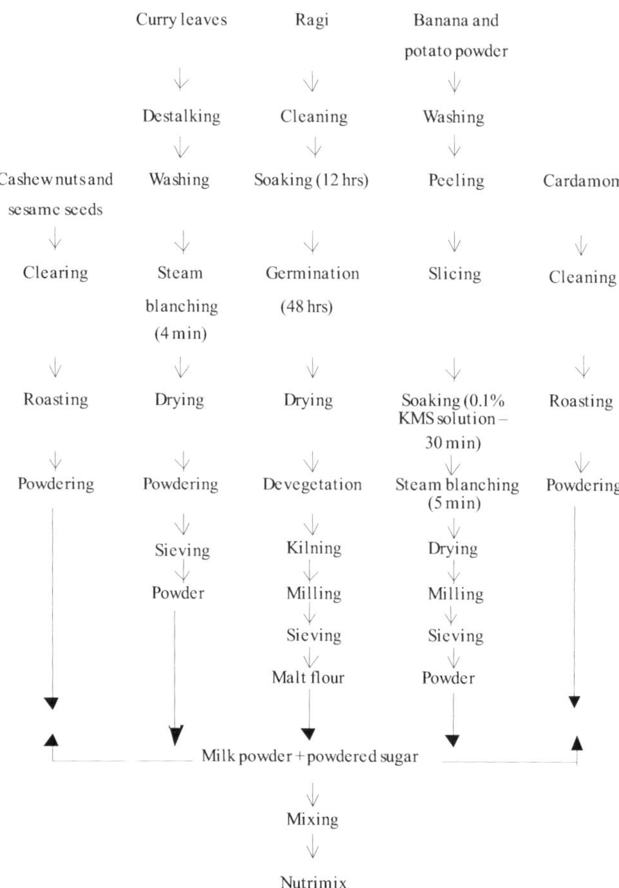

Figure 4: Flow diagram of ragi Nutrimix

Value Added Products of Pearl Millet

The Department of Food and Nutrition has also developed as many as 33 value added products such as *nankhatai*, cake, sweet biscuits, sweet salty biscuits, pasta, fryums, weaning mixture, *gatta* curry, *idli, burphi*, sprouted *chat, halwa, ladoo* etc. of the maize varieties/hybrids and also several high value products of pearl millet which can be popularized as these products are highly nutritive.

Processing of Pearl Millet for Value Addition

For any food processing unit the first and foremost requirement is raw material processing by sun drying, screening for cleaning of foreign material like sand particles, agricultural waste if any, followed by mechanical processing including dehusking, grinding, milling - converting into flour or fine particles. Later it needs to be converted into intermediate stage to bring the final precuts. Conversion is carried by addition of other ingredients, flavors, milk; water. The mechanized system like mixers, kneaders, baking, cooking can also be employed. Finished products needs to be tested for its quality and other parameters followed by packing it in containers either in ready to sell form like pouches, pet jars and others or it can be preserved for final packing as and when required.

Key Goal for Food Units

- Taste and Quality final product
- Shelf life of the product
- Hygienic working condition during production
- Cost effective and affordability

Precautions for Food Processing Units

- No adulteration
- Less damage during packaging and transposition
- Meeting of food safety standard as per government norms

Table 7: List of machineries required for millets milling units

S.No	Name of equipments	Capacity	Quantity
1	Flour Mill	50 kg/ hr	1
2	Auto seiver	50 kg/hr	1
3	Flour blenders	100 kg/hr	1
4	Weighing scales Heavy	Heavy-1Light-1Small-1	1
5	Sealers (vacuum)	——————-	1
6	Storage Bin (raw, Intermediate products)	500, 100, 50 and 25 kg	4

Extrusion Technology

Extrusion is a process that combines several unit operations, including mixing, kneading, shearing, heating, cooling, shaping and forming. It involves compressing and working raw materials e.g. flours, starches, proteins, salt, sugar and other minor ingredients, to form a semi-solid mass under a variety of controlled conditions and, then forcing into to pass through a restricted opening such as a shaped hole or slot at a predetermined rate. Heat is applied directly by steam injection or indirectly through a heated barrel or the conversion of mechanical energy. The final process temperature in the cooking extruder can be as high as 200°C but the residence time is relatively short i.e. 10-60 seconds. Thus extrusion cooking is also called a high temperature short time (HTST) process. This process of HTST, extrusion bring gelatinisation of starch, denaturation of proteins, modification of lipids and inactivation of enzymes, microbes and many antinutritional factors.

Extrusion Application in Food Industry

Infant foods, instant beverage, cooked whole grains, quick cooking vegetable, breakfast cereals, bread, sticks, pasta products, snack foods, meat analogues, pet food, animal feed, confectionery, textured vegetable proteins, biscuits, food bars, and meat products.

Parts of the Extruder

The equipment used for extrusion cooking is known as extruder. The main parts of the extruder are drive mechanism, feed hopper, barrel, and screw, die etc.

The central part of a food extruder is the screw. Typical screw can be divided into three sections. The section of the screw, where the feed enters is commonly called the feed section and normally has deeper flights or flights of greater pitch. The section of the screw, where the feed material is compacted and converted from a flowing granular sticky mass to relatively uniform plasticized dough is called the transition (or) compression section. After the material leaves the transition section and before it enters a die, a mattering section of the screw with relatively shallow flights for increased restrictions of the channel area is used to increase the temperature of the material for cooking.

Equipment

The basis of extrusion cooking is a screw pump using twin or single screw. The single screw extruder is still used in industry and its origins are in pasta manufacture, which is a low temperature, high moisture process. The single screw extruder has a wide residence time distribution and is generally a poor mixer. The single screw extruder may be operated with a pre-conditioner, which may be used to add water either directly or as steam. There is nonetheless, considerable interest in developing the single screw extruder as a modular design with mixing elements, such as the cavity transfer mixer.

A greater variety of twin screw devices is available with intermeshing (or) non-inter meshing Co (or) counter rotating extruders. The counter rotating intermeshing twin screw extruder is cited for use with low viscosity materials and generally operates at lower screw speeds than its co-rotating relative. However, the most commonly employed type is the intermeshing, co-rotating extruder comprising self-wiping, conveying and reversing elements, orifice discs and mixing paddles, which may be assembled to convey forward or backward. Liquids and other solid or semi-solid ingredients are often fed into the extruder at different stages along the barrel length.

The twin screw extruder has the capability for good mixing and narrow residence time distributions. The versatile design of the screw means that the mixing stages can be synchronized with the temperature zones and the pressure may be designed to follow a complex history involving a section at atmospheric pressure in order to vent the extruder or to inject materials or gases. The most recent development in extrusion technology is to combine it with pressures. The combined "supercritical fluid extrusion' process introduces the supercritical fluid, usually carbon dioxide, carrying soluble micronutrients, flavours and colours, into the melt stage of the extruder. It has been demonstrated for breakfast cereals, pasta and confectionary production.

A number of innovations are possible as the material leaves the screw section and passes through the die. Multiple dies may be designed with geometries to reduce pressure more or less rapidly to atmospheric. Co-extrusion in which filler is pumped into the centre of a hollow extruder has been popular for a number of years and has enhanced the product types available through this technology. The occurrence of flow defects and the visco- elastic nature of food materials dictate that the material exiting the extruder does not conform to the shape of the die. The loss of super heated water as steam has been the basis of many novel extruded products and in these cases, the extruded diameter is markedly greater than the die section. At its simplest, extrusion cooking may be used as a single stage process to produce pasta or expanded cut pieces. However, most extrusion technology is used with other downstream processes for rolling, crimping, cutting, toasting and coating.

Further technological innovations are possible with the twin screw extruder in handling wet materials. The extruder may be used for multiple foods such as dehydrated potato and potato mash. Similarly, nuts and fruit slurries may be added at appropriate points to avoid high shear or high temperature.

Development Process of Extruded Products

A. Raw Material Preparation

Raw material preparation for processing into extruded- cooked consumption products depends, first of all, its constituents. Depending on their quality, they have to be comminuted, weighed and proportioned according to the recipe as well as mixed prior to pouring them into the extruder – cooker. If conditioning is required, a little water is added prior to mixing.

B. Extruded – Cooking Process

The extruded – cooking process is taking in the final zone of the apparatus. As a result of combined effect of the temperature (120-200°C) and pressure essential physico-chemical changes occur in the material while extrusion – cooking. When leaving the die block aperture, the material rapidly expands and the structure of the obtained products is very much like that of a honey-comb, the structure being shaped by bundles of fused protein fibres.

C. Shaping, Drying and Packing

Depending on actual needs and application i.e. whether the extruded – cooked material is to be used as a semi-finished product or a finished products, the material leaving the extrusion-cooker is subjected to respective shaping. The material is shaped by the die. The applications of the various kinds of die shapes enable the obtaining of many different forms of products like balls, rings, stars, alphabet letters etc. The length of the product is adjusted by regulation of the speed rate of rotary knife which is installed outside the nozzle.

A consecutive production step is subjecting the obtained material to drying, usually to about 90% of the humidity content and then to chilling. A tunnel dryer is usually used for its drying, the drying medium being air heated by gas or by steam to about 100°C. For chilling, air of an ambient temperature (15-20°C) passes through the apertures of the perforated tunnel dryer belt. The last step of the production protein starch extruded – cooked food stuffs, is their packing into plastic foil or paper bags. The storage life of extruded – cooked products is considerable and under correct storage conditions they can be stored for many months without any loss of their sensory or nutritive proteins.

Macaroni Products

Macaroni products also termed as pasta products include macaroni, spaghetti, vermicelli and noodles. Macaroni is a formed hollow tubes, spaghetti is a small solid rods, vermicelli is a tiny rod and noodles are flat strips. Pasta products are also available in the shapes of shells and alphabets. Pasta products are made from coarser durum wheat except noodles are made from wheat flour. Eggs are often added to the dough used in the manufacture of macaroni products. Other materials used are groundnut meal, soya flour and cassava flour. The preparation can be enriched by addition of vitamin and amino acids.In the manufacture of macaroni, the semolina / flour is made into a stiff dough using 25-30% of water on weight at 32-38°C. After a rest period, the dough is kneaded at about 30°C in a cylindrical machine, equipped with helical blades. The dough is then extruded at high pressures through appropriate dies to make tubes or strap shaped products. Heavy pressure ensures that the product is translucent and free from air bubbles. Finally the product is dried at controlled temperatures and humidity and cut into desired lengths. The extruded pastes must be dried from about 30% moisture to about 12 % before packaging.

Development of Vermicelli and Noodles

Ingredients Required

Maida flour	:	100 gm
Salt	:	2 gm
Water	:	35 ml

The maida flour was steamed for 5 minutes and conditioned by adding the required quantity of water and salt. The mixture was fed into the extruder by selecting the die (vermicelli and noodles). The dough was extruded. The extruded vermicelli and noodles were steam cooked for 5-10 minutes and dried at 60°C for 6 hrs in a cabinet drier, cooled and packed in polyethylene pouches.

Development of Spaghetti and Macarini

Ingredients Required:

Maida flour	:	100 gm
Salt	:	2 gm
Water	:	35 ml

The maida flour was steamed for 5 minutes and conditioned by adding the required quantity of water and salt. The mixture was fed into the extruder by selecting the die (spaghetti and macaroni). The dough was extruded. The extruded spaghetti and macaroni were steam cooked for 5-10 minutes and dried at 60°C for 6 hrs in a cabinet drier, cooled and packed in polyethylene pouches.

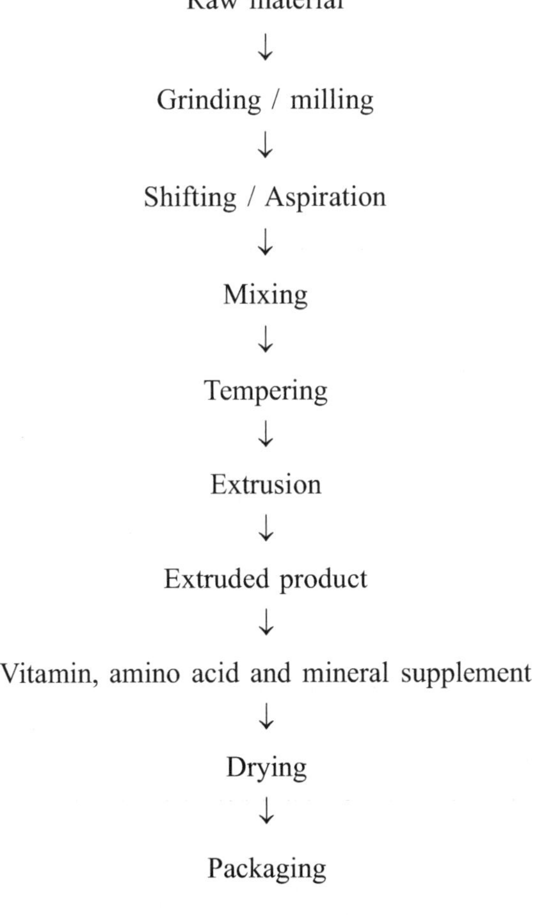

Fig 5: Flow diagram of extruded products

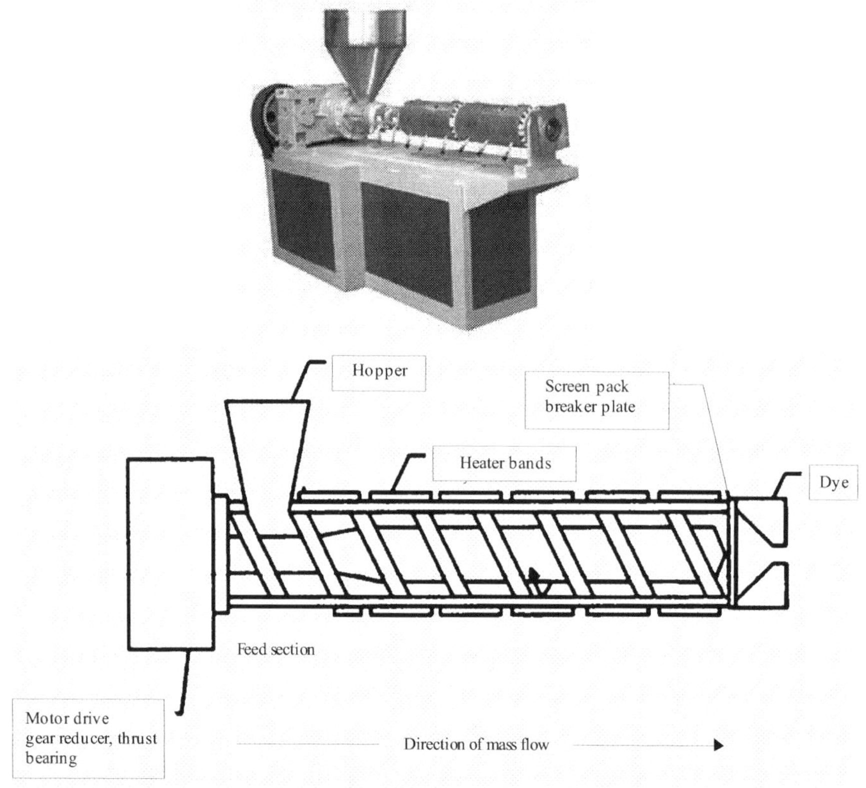

Figure 6: Single screw extruder

Preparation of Nutrient Enriched Extruded Products

Ingredients

Wheat flour	:	80 gm
Defatted soya flour	:	20 gm
Water	:	32 ml
Salt	:	2 gm

The wheat flour and defatted soya flour was mixed thoroughly. The flour was steamed for 5 minutes and conditioned by adding the required quantity of water and salt. The dough was prepared and extruded through an extruder. The noodles were steam cooked for 5-10 minutes and dried at 60°C for 6 hours in a cabinet drier. The noodles were cooled and packed in polyethylene pouches.

Development of Weaning Food/ Supplementary Food

There is a need for nutritious weaning foods that are acceptable and affordable to low income populations. Guidelines state that an ideal weaning food must be nutrient dense, easily digestible, of suitable consistency and affordable to the target market. The inclusion of various cereals and legumes in the development of infant food supplements has been extensively investigated. Cereals, the main source of calories and, are adequate in methionine and cysteine and are a good source of B-complex vitamins but limiting in lysine.

Most legumes are rich in lysine but low in sulfur amino acids, allowing the combination of cereals and legumes into complementary blends at about a 50:50 ratio of the two proteins Though simple household processes are advocated in the development of nutritious weaning food mixtures targeted at low income populations, the safety and convenience of the final product is also very important. Low moisture products that may be stored safely in the absence of refrigeration and reconstituted quickly on demand may reduce the need for infant foods to be prepared and stored at ambient temperature over extended periods and the associated gastrointestinal infections that occur due to microbial contamination.

Low-cost extrusion cooking is a versatile food processing technology that rapidly mixes and kneads feed material at temperatures of over 100°C to cook and dry the product in a relatively short time. This thermal process improves the nutritional quality of the raw food material and eliminates vegetative microorganisms. The result is a cooked extrudate that can be milled into flour and then reconstituted into a pap for infant feeding. In this study, simple indigenous techniques of food processing such as decorticating and toasting were combined with low-cost extrusion to create an experimental product that could be produced on a large scale as a community-based, sustained effort. The nutrient profiles of the formulated blends were evaluated by chemical assay.

Under ICDS Supplementary feeding is provided for 300 days in a year by providing Supplementary feeding, the Anganwadi attempts to Bridge the caloric gap between the national recommended and average intake of children and women.This pattern of feeding aims only at supplementing and not substituting for family food. The types of food varies from state to state, but usually consists of a hot meal cooked at the anganwadi, containing a varied combination of pulses, cereals, oils, vegetables and sugar/iodised salt. Some states provide a ready to eat meal, containing the same basic ingredients. There is flexibility in the selection of food items, to respond to local needs. 100grams of weaning food provides 350 calories and 8.5grams protein.

Table 8 Nutrition details of weaning food/100 gram

Category	Quantum of weaning food provided	Present cost per beneficiary per day	Protein (gms)	Energy (Kcal)	Vitamins & Minerals (% of RDA)
Children 6-36 Months	130	Rs.4.00	11	455	50%
Children 6-36 Months (Severely malnourished)	190	Rs.6.00	16	665	50%
Pregnant Women and Nursing Mothers	160	Rs.5.00	13.5	560	50%

Breakfast Cereals

Important products commercially produced and marketed as breakfast cereals include Dalia (Farina), rolled oats and corn and wheat flakes. While dalia and rolled oats require cooking before use, the corn and wheat flakes are ready-to-use. In India, use of breakfast cereals, however, is limited only to affluent sections of society because of their relatively high cost.

(a) Dalia: Dalia or farina is nothing but the middling fraction obtained during roller flour milling. These are chunks of wheat endosperm free of bran and germ. In the manufacture of dalia, it is beneficial to use hard wheat since dalia milled from soft wheat becomes excessively pasty upon cooking. An instant porridge based on precooked wheat dalia and milk solids has been developed for astronauts.

(b) Rolled oats: In the manufacture of rolled oats, the cleaned grains are heated to 100°C for one hour. This reduces the moisture to about 6% and partially dextrinizes the starch. In addition, the roasting step renders the hulls fragile which are easily removed by passing through a pair of circular stones. The dehulled grains (groats) may be flaked directly or these may be first cut into pieces by rotary granulators. Thinner the flake, shorter the time takes for cooking. The quick oats are processed from ½ to ¼ the size of the whole groats and these require about and these require about 5 minutes boiling.

(c) Cereal flakes: Flaking is relatively a simple process consisting of cooking whole or fragments of cereal grains, flattening the tempered grains by passing through rollers and toasting the resultant flakes at high temperature to impart caramelized flavour.

Both wheat and corn flakes are commercially produced in India; but total production is limited because of relatively high price and limited consumer demand of this product. The two most encountered problems are excessive breakage during transportation and oxidative off flavours generally associated with the commercially available samples released to industry; but there is a tremendous potential for its commercial production in view of the large production of sorghum (10.5 million tonnes) in our country and there is no industrial value-added product produced from it at present

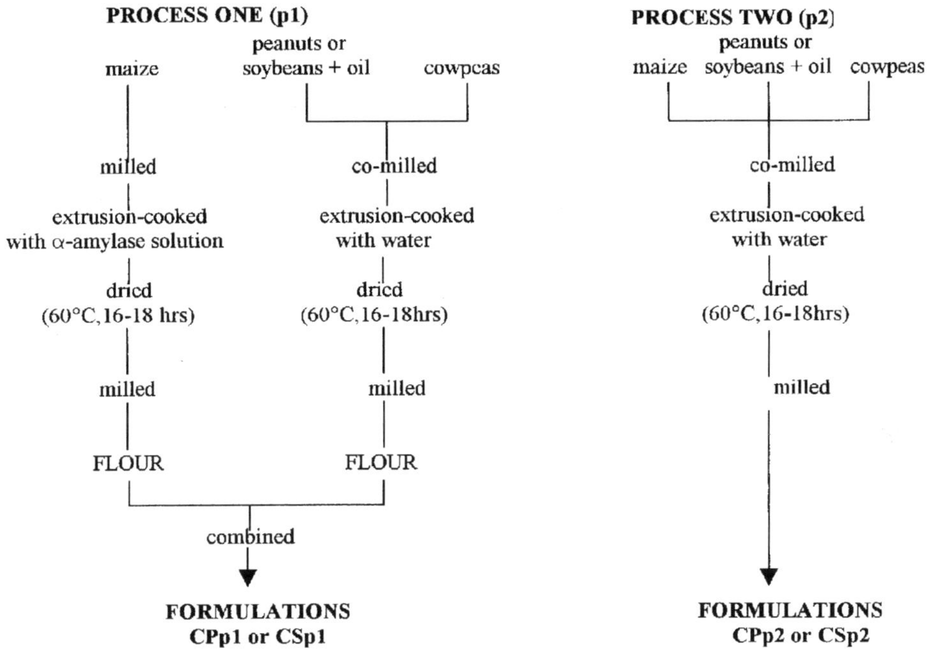

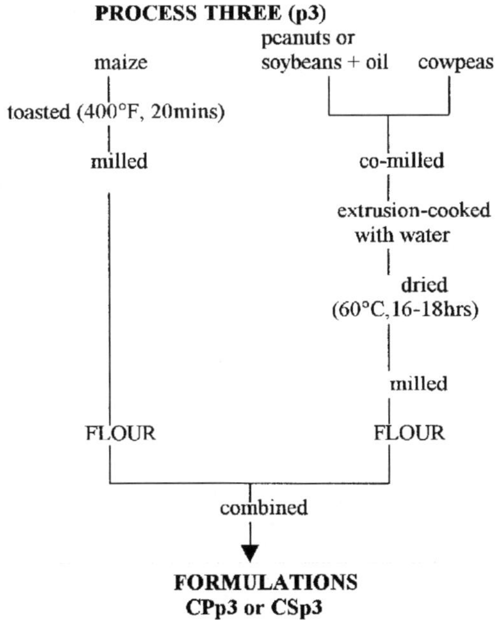

Figure 7: Process flow diagram of weaning food preparation

Recently, breakfast cereal flakes from sorghum have been developed at DFRL, Mysore. The product is crisp and reconstitutes well in hot milk. The process know-how is yet to be

Preparation of Extruded Products

Preparation of Noodles

Ingredients

Maida flour	:	800 gm
Mixed pulse flour	:	200 gm
Salt	:	2 g/ 100 gm

Mixed Pulse Flour

The ground pulse flour (Bengal gram, green gram and peas) were mixed in the proportion of 1:1:1 to prepare mixed pulse flour.

Method

1. The maida flour was steamed separately for 10 min in the idli cooker.
2. The mixed pulse flour was also steamed for 15 min in the idli cooker.
3. Then the steamed flour was cooled and sieved again through the BS 60 sieve and mixed (maida + pulse flour + salt).
4. The above prepared mix was put into an automatic extruder.
5. Water was added (400 ml/kg) to the mix.
6. The kneading was continued for 15-20 min and extruded into noodles (0.8 mm dia).
7. The prepared noodles were steamed for 12 min in the idli cooker for partial gelatinisation.
8. Then tempered at room temperature for overnight (10 hr) and dried in the cabinet direr at 50°C for 2-4 hr.

Preparation of Savoury Products

A. Uppuma: Ingredients Needed

Extruded product	:	100 gm
Onion	:	15 gm
Green chillies	:	3-4 nos.

Curry leaves	:	Few
Mustard	:	½ tsp,
Bengal gram dhal	:	½ tsp,
Blackgram dhal	:	1 tsp.
Salt	:	3 gm
Oil	:	20 ml
Water	:	250 ml

Method: The oil was heated in a frying pan and the mustard was added and allowed to crack. Then the black gram and bengal gram dhal, onion, green chillies and curry leaves were added to it. When the onion turned to golden yellow colour, required amount of water was added and boiled. Extruded product and salt were added to the boiling water and stirred. Cooking was continued till the sample became soft.

B. Curd Bath: Ingredients Needed

Extruded product	:	100 gm
Mustard	:	2 gm
Green chillies	:	4 nos
Ginger	:	2 gm
Curry leaves	:	2 gm
Curd	:	100 ml
Oil	:	10 ml

Method: Extruded product was steam cooked and kept aside. Green chillies and ginger were cut into pieces. Seasoning was done using oil, mustard, green chillies and curry leaves and mixed with the curd. The cooked sample was added to curd, mixed well and served.

C. Tomato Bath: Ingredients Needed

Extruded product	:	100 gm
Tomato	:	25 gm
Green chillies	:	4 nos.
Curry leaves	:	2 gm
Onion	:	20 gm
Spices (clove, cardamom and cinnamon)	:	2 gm

Method: Sample was steam cooked and kept aside. To the heated oil the spices and finely chopped onion, green chillies, tomato and salt was added and continued cooking for 5 minutes. The cooked sample was added and mixed well.

D. Lime Bath: Ingredients Needed

Sample	:	100 gm
Lime juice	:	25ml
Chopped onion	:	20 gm
Red chillies	:	5 gm
Chopped ginger	:	2 gm
Bengal gram dhal	:	5 gm
Mustard	:	2 gm
Turmeric powder	:	2 gm
Oil	:	10 ml
Salt	:	To taste

Method: Cooked sample was seasoned with mustard, bengal gram dhall and chillies. Lime juice, salt and turmeric powder were added and mixed well.

New Product Development (Dough and Batter)

Dough and batter are made from a combination of flour, fat, and water or milk. By including other ingredients such as sugar, yeast, fruit, and nuts there is plenty of scope to create a variety of products.

The major difference between dough and batter are the moisture content of a dough is lower than that of a batter, and consequently dough has a heavier consistency and can be moulded into many shapes. Both dough and batters can be fermented to produce many products. For example fermented maize products such as 'Kenkey' in Ghana and 'Bagone' in Botswana are considered to be staple foodstuffs.

The production of batters and dough involves mixing ingredients together to a smooth and uniform consistency. This can be done manually or with a powered mixer. The type of powered mixer required will depend upon the product being prepared. For example, a batter will need a balloon whisk-type mixer whereas dough will require a hook-type mixer. There are general purpose mixers available for use on a small scale, which have attachments for both products. There are various ways in which dough and batters may be subsequently processed.

Noodles can be made from a range of flours including rice, wheat, maize, and potato. The dough can be processed in one of two ways, either rolled out into thin sheets of dough and cut into strands, or extruded.

Preparation of Noodles

The strands are then steamed, and may either be eaten fresh, or processed further by drying. Drying can be achieved using either a solar or a fuel-fired dryer. It is possible to dry the noodles in the sun, but the quality of the finished product is likely to be lower. Packaging requirements for dried noodles include a moisture-proof package (e.g. polythene) and an outer carton/box to prevent crushing.

Table 9: Production stages of rice noodles

Ingredients	Processing stage	Equipment
Rice flour, water	Mix ingredients to a dough	Powered mixer (optional)
	Roll into thin sheets of dough	Rolling equipment
	Cut into strands	Cutting equipment
	Extrude	Extruder
	Steam	Steam blancher
	Dry	Solar dryer
		Fuel-fired dryer
		Electric dryer
	Pack	Polythene sealing machine

Porridge-Type Products

Flours and flakes are often cooked with water to produce a porridge. These porridges are prepared in the home as a main meal, and their consumption is prominent in many parts of Africa (e.g. foodstuffs such as 'banku' and 'ugali' made from maize and consumed in Western and Eastern Africa respectively).

Weaning foods are a combination of cereals and pulses prepared as a mixture of flours and flakes. These foods are designed to be made into a porridge by adding water or milk. Many weaning foods are made by multinational manufacturers on a large scale. However, it is possible to produce a weaning food on a small scale by:

- Preparing the correct combination of raw materials
- Roasting the mixture of cereals and pulses
- Milling using a hammer mill
- Packing as a dry mix.

Table 10: Production stages of tortilla

Ingredients	Processing stage	Equipment
Maize, lime*	Boil grains in water and lime (until the grains can be peeled by hand)	Boiling pan
	Wash the maize	Clean potable water
	Mill grains with the addition of water	Plate mill
	Mix to a dough with water	Powered mixer (optional)
	Shape	Rolling equipment
	Fry on a hotplate	Hotplate
	Consume immediately or pack	Heat sealing machine

*The lime involved is not the fruit but the chemical.

Such mixtures can then be reconstituted in the home and mixed to a porridge-like consistency with water. It is vital that nutritional advice is sought to ensure that the mixture is nutritionally balanced, containing sufficient energy, protein, and other nutrients which will maintain the health and growth of the child. Such foods are vital since it is in this transition period from the breast to the family pot that the child is most likely to suffer from protein-energy malnutrition and other nutritional deficiency diseases.

Physical Characteristics of Extruded Products

Method

A. Physical Characteristics of Uncooked Noodles

The appearance, colour, flavour and texture of the uncooked (raw) extruded products were observed and recorded.

B. Breaking Strength of Extruded Products

To study the breaking strength of extruded products a simple technique was developed. An extruded product strip of 6.0 cm length was taken and its ends were tied horizontally to supporting stands about 15 cm above the ground level. A known weight (5,10 and 15 g) was dropped on the tied strip at different height. When the extruded strip breaks the breaking strength of the sample was recorded as x gm for y cm. 3. Rehydration characteristics of the extruded product

$$\text{Rehydration ratio} \quad \frac{\text{Drained weight to rehydrated sample}}{\text{Weight of the dehydrated sample}}$$

C. Cooking Quality

Method

The parameters such as cooking time, cooked volume, water absorption and solid loss were studied by using cooking quality test. A sample of 5 gm was taken in a boiling test tube (18.5 cm length and 1.4" dia) and its initial volume was measured. Fifty ml of water was added to it and the test tube was placed inside a boiling water bath. The cooking time was noted when the water inside the test tube started boiling. A few strips were periodically withdrawn and pressed between two glass slides to find out the complete doneness of the noodles. The excess water from the cooked sample was discarded and the weight and the volume of the same were noted.

The water absorption was calculated by determining the weight of the cooked product and the result was expressed in per cent in relation to the uncooked product. The weight and volume of the cooked samples were expressed as gm and ml per 100 gm of uncooked product respectively. The drained water was taken in a tarred petridish and dried in oven and the difference gave the percentage solid loss in cooked samples.

D. Analysis Procedure

The total protein content of the soya incorporated noodles was estimated by using micro kjeldhal method (AOAC, 1975).

Hundred milligram of sample was transferred into the digestion flask along with two gram of potassium sulphate and 80 mg of mercuric oxide as catalyst and 3 ml of Conc.H_2SO_4 was added and digested until the solution becomes colourless.

The digested sample was transferred into a 100 ml volumetric flask and made up with distilled water. A known aliquot was transferred into the distillation flask and distilled. Five ml of 4% boric acid with few drops of mixed indicator was placed under the condenser, with the condenser tube extending below the boric acid solution. Eight ml of NaOH sodium thiosulphate solution was added to the sample. The sample was steam distilled for 8 minutes to collect the ammonia from the sample into the boric acid.

This was titrated using 0.01 N HCl till the disappearance of blue colour. Simultaneously a blank was also carried out without the sample. The nitrogen content was calculated as gm/kg and by using the conversion factor 5.95 to protein content of the sample and expressed as gm / 100 gm.

Wheat Based Foods

Type of Milling

Grains and pulses are milled to produce flours which are used to produce many types of food. Three main types of mill are available:

- Plate mill
- Roller mill
- Hammer mill.

The choice of mill will depend on the raw material and also on the envisaged scale of production, a range of different mills, but for details of the milling process and in-depth guidelines relating to choice of machinery, Flours and flakes can be used as starting materials for the following types of product.

Wurld wheat prepared by lye peeling and acetic acid bleaching can also be expanded considerably by equilibrating to 20% moisture and then drying in a spouted bed dryer. In this dryer, air at 400-500°F and at 100 ft/sec carries the heat to the partially fluidized bed of grain. Expanded wurld wheat takes much shorter time for cooking compared to wurld wheat.

At present biggest constraint in upgrading the quality and production of expanded cereals seems to be the non-availability of a suitable low cost mechanized roaster and lack of suitable packaging for retail marketing. The present method of production is laborious and not easily amendable to large scale production and the product is liable to contamination under retail marketing and therefore does not create enough confidence of quality product among consumers.

Principles of Baking

Wheat flours find their principal applications in the production of bakery products. Most bakery products-unlike other wheat products such as alimentary pastes or noodles and unpuffed breakfast cereals- are leavened; that is, they are raised to yield baked goods of low density. The term baking strictly refers only to the operation of heating dough products in an oven. But since there are many steps that must take place before the oven if baking is to be successful, baking has come to mean all of the science and technology that must precede the oven as well as the oven-heating step itself. We will consider baking in this broader sense.

Although there are a great many bakery products which grade one into another in term of their formulas, methods of preparation, and product characteristics, it is possible to classify bakery products according to the way in which they are leavened. This classification, though not perfect, is useful. Four categories may be defined;

- *Yeast –raised goods* include breads and sweet doughs leavened by carbon dioxide from yeast fermentation
- *Chemically leavened goods* include layer cakes, doughnuts, and biscuits raised by carbon dioxide from baking powders and chemic agents
- *Air- leavened goods* include angel cakes and sponge cakes made without baking powder

- *Partially leavened goods* include pie crusts, certain crackers, and other items where no intentional leavening agents are used yet a slight leavening occurs from expanding steam and other gases during the oven-baking operation.

Production

Baking is a process which uses heated air to alter the eating quality of foods. A secondary purpose is preservation, by destroying microorganisms and reducing the moisture content at the surface of the food.

In this chapter, 'baked goods' refers to breads and flour confectionery, but it is also possible to apply the process to meat, fruits, and vegetables. The process in these cases is often referred to as 'roasting'.

Baked goods are produced from either doughs or batters, which are a mixture of flour and water made by mixing, beating, kneading, or folding. The process will depend on the product being made and the ingredients used.

The tables overleaf outline the types of process and the equipment required to make a representative range of products.

Processing

(a) Raw materials: Wheat flour contains a substance called gluten. This is responsible for providing the internal structure of products such as bread and some biscuits. Depending on the variety and the climate in which it is grown, wheat will contain different quantities of gluten. For example, flours which contain low levels are known as 'soft' or 'week' flours, and those with high levels are known as 'strong' flours. The type of wheat flour used will depend on the product being made. Bread and biscuits require a strong flour, whereas cake flours are usually medium to weak in strength.

Table 11: Processing stages of baked products

Process/product	Mix	Ferment	Form	Prove	Knock back	Bake	Pack
Leavened bread	*	*	*	*	*	*	*
Buns	*	*	*	*	*	*	*
Biscuits	*		*			*	*
Pastries	*		*			*	*
Cake	*		*			*	*
Unleavened bread	*		*			*	*

Flours from cassava and rice may be used as a substitute for wheat, but as they do not contain gluten, a maximum substitution rate of 25 per cent is possible before differences in taste and texture become very noticeable. Additionally, there are many other products that are made completely from flours such as maize or rice.

(b) Dough Mixing: All doughs and batters are mixed to achieve a smooth consistency and to ensure an even distribution of ingredients. This is usually done using simple hand-operated tools, but if a large quantity is to be produced it may be more convenient to use a powered mixer.

Mixers are classified according to the type of product that they are used for, i.e. either solids or liquids. A common solids mixer operates with a planetary action and can be fitted with a hook for pastry or dough, or a 'K-beater' for dry powders. Liquid mixers include a balloon-whisk attachment for batters or propeller-type mixers for thin liquids such as sugar solutions.

(c) Fermentation: For some products it is not necessary for the dough to rise, or for the final texture to be open and porous. Such examples are unleavened breads such as roti, chapatti, and tacos, which as the group-name suggests do not require fermentation. Other products, such as scones or soda bread, rely on gas produced by added sodium bicarbonate to raise the dough and produce the desired crumb texture. In cakes, air is incorporated into the batter during mixing to give an open structure to the crumb after baking.

Table 12: Equipments required for baked products

Processing stage	Equipment
Mix	Liquid mixer
	Solid mixer
Ferment	Proofing cabinet
Form	**Bread:**
	Dough moulder
	Biscuits:
	Rolling machinery
	Depositor
	Dough cutter
	Pastry:
	Rolling machinery
	Pastry mould
	Cakes:
	Baking tints
Prove	Same as for the fermenting stage
Knock back	
Bake	Oven
Pack	Polyethylene or cellophane wrapping/sealing machines

Only bread and buns are fermented to produce the open porous crumb. In this case added yeast ferments (breaks down) the sugars in the flour (and/or added sugar) to produce a gas, carbon dioxide, which raises the dough. No special equipment is required to bring about fermentation. The dough is covered with a damp cloth or polythene and left to rest at a temperature of 32-35°C until it becomes fully inflated with carbon dioxide. To achieve this, the dough may be left near to the oven where it is warm, or in a temperature-controlled cabinet known as a proving cabinet.

Proving is the name given to a process of secondary fermentation which is applied to bread and bun dough. It is necessary because as the dough is being shaped some of the carbon dioxide is lost and the dough structure partially collapses. The aim of proving is to ferment the dough a little more and regain the carbon dioxide levels and therefore the open structure.

(d) Kneading/forming/ moulding: Kneading is a process of stretching and folding dough. In doing so the gluten fibers are stretched and the consistency of the dough becomes smooth. Kneading is most usually done by hand, but if a large quantity is being produced it can be a tiring task and a powered kneader may be preferred.

After kneading, the dough has to be formed into the desired shape and size. The method for doing this will depend upon the type of product, for example, bread dough can be cut, folded, or plaited into shape. For production above the household level, a machine known as a dough divider can be used to cut the dough into accurately-weighed pieces, out of which the individual loaves or buns will be produced.

Biscuit doughs are rolled out and cut using either a knife or a shaped biscuit cutter. Pastry dough can be handled in a similar way. In the production of pies, a pastry mould is used to retain the shape of the dough during baking.

(e) Baking: During baking, food is heated by the hot air, and also by the oven floor and trays. Moisture at the surface of the food is evaporated by the hot air, and this leads to a dry crust in products such as bread and many biscuits. If a glazed product is desired then it will be necessary to either inject steam, or place a pan of water in the oven. Oven temperatures for baking depend on the type, size, and shape of the product, and the nature of the ingredients.

Moulding Dough

Table 13: Oven temperatures for different baking products

Product	Degrees F	Degrees C	Time (minutes)
Cake	350	176	45-60
Pastry	450	232	15-20
Biscuits	425-450	218-232	10-15
Bread	400	204	30-40

A range of wood, charcoal, gas, or electric ovens is available. The choice will depend on the cheapest and most widely available fuel in a particular area.

This classification refers to the intended source of the leavening gas; however, the intended source is not the only source of leavening. Leavening gas can produce leavening only if it is trapped in a system that will hold the gas and expand along with the gas. Therefore, much of cereal science related to baking technology is really the engineering of food structures through the formation of correct doughs and batters to trap leavening gases, and then the coagulation or fixing of these structures by the application of heat. This brings in the need to understand further some of the properties of flour and certain other baking ingredients.

Fermented Products

Idli and Dosai are the two most popular fermented products used either as snack or in breakfast, especially in Southern parts of our country, and are prepared from fermented batter prepared from rice and black gram. At present, these are prepared mostly in homes or marketed by food service establishments. There have been some attempts to preserve idli in ready-to-eat state by thermal processing in cans and retort pouches. However, this concept has not caught up and there is hardly any demand for thermally processed idlis. Secondly, attempts have also been made to develop instant mixes for idli and dosai. The concept of instant mixes is quite attractive and a few firms are already marketing them. However, idlis and dosai prepared from these mixes are not soft and fluffy and also lack fermented flavour. Some attempts have also been made to characterize the microorganisms responsible for fermentation of idli batter. Surprisingly, even after characterization of organisms involved in fermentation, no attempt has been made to develop instant mixes having lyophilized or micro-encapsulated cultures.

This aspect definitely needs more attention. Thirdly, in fermented batters surface active proteins and polysaccharides of black gram are well conditioned to retain large volume of gases to give soft fluffy texture. In the ready mixes, this conditioning does not take place and additional hydrocolloids may help in providing support to gel structure in retaining gases generated by the action of chemical leavening. This aspect also needs careful scrutiny for improving the quality of mixes.

Standardization and Preparation of Idli Ready Mix

Idli is a small, white acid – leavened, and steamed cake made by bacterial fermentation (12-18 hours) of a thick batter made from carefully washed rice and dehulled blackgram dhal. The idli cakes are soft, moist and spongy had a desirable sour flavour. For idli, the rice was coarsely ground and the black gram was finely ground.

The soft spongy texture observed in the leavened steamed idli made out of black gram is due to the presence of two components, namely surface active protein (globulin) and an arbinogalactan (polysaccharide) in black gram. The mucilaginous principle of blackgram is identified as arbinogalactan. It is believed that this mucilaginous principle helps in the retention of carbondioxide during the fermentation of the thick batter and is thus responsible for the soft spongy honey comb texture of the idli. Fermentation brings about physical and chemical changes in the idli batter. With the progress of fermentation there is an increase in batter volume, acidity and non protein nitrogen.

Instant idli mixes eliminate the traditional method of grinding of both the ingredients and the leavening is produced by the action of chemical leavening agents.

Procedure Preparation of Rice and Black Gram Flours

The parboiled rice IR 20 was soaked in water for 5 hours. Then the water was drained completely and dried in solar drier for 5 hrs. The dried rice was ground in a mixie and sieved through BS 36 sieve. Black gram dhal was also ground in a mixie and sieved (BS 36). The rice and black gram flour were dried in a cabinet drier at 80°C for 2 hrs, cooled and packed in air tight containers till they were used.

Materials Required

Rice flour	:	80 g
Black gram dhal flour	:	20 g
Salt	:	2 g
Citric acid	:	0.05 %
$NaHCO_3$	:	0.2 %

Method

1. The ground rice and blackgram dhal flours were mixed at the ratio of 4:1.
2. The salt, citric acid and sodium bicarbonate were added and mixed thoroughly.
3. Another batch of instant mix was prepared by adding yeast (0.1%) without the addition of citric acid and sodium bicarbonate.

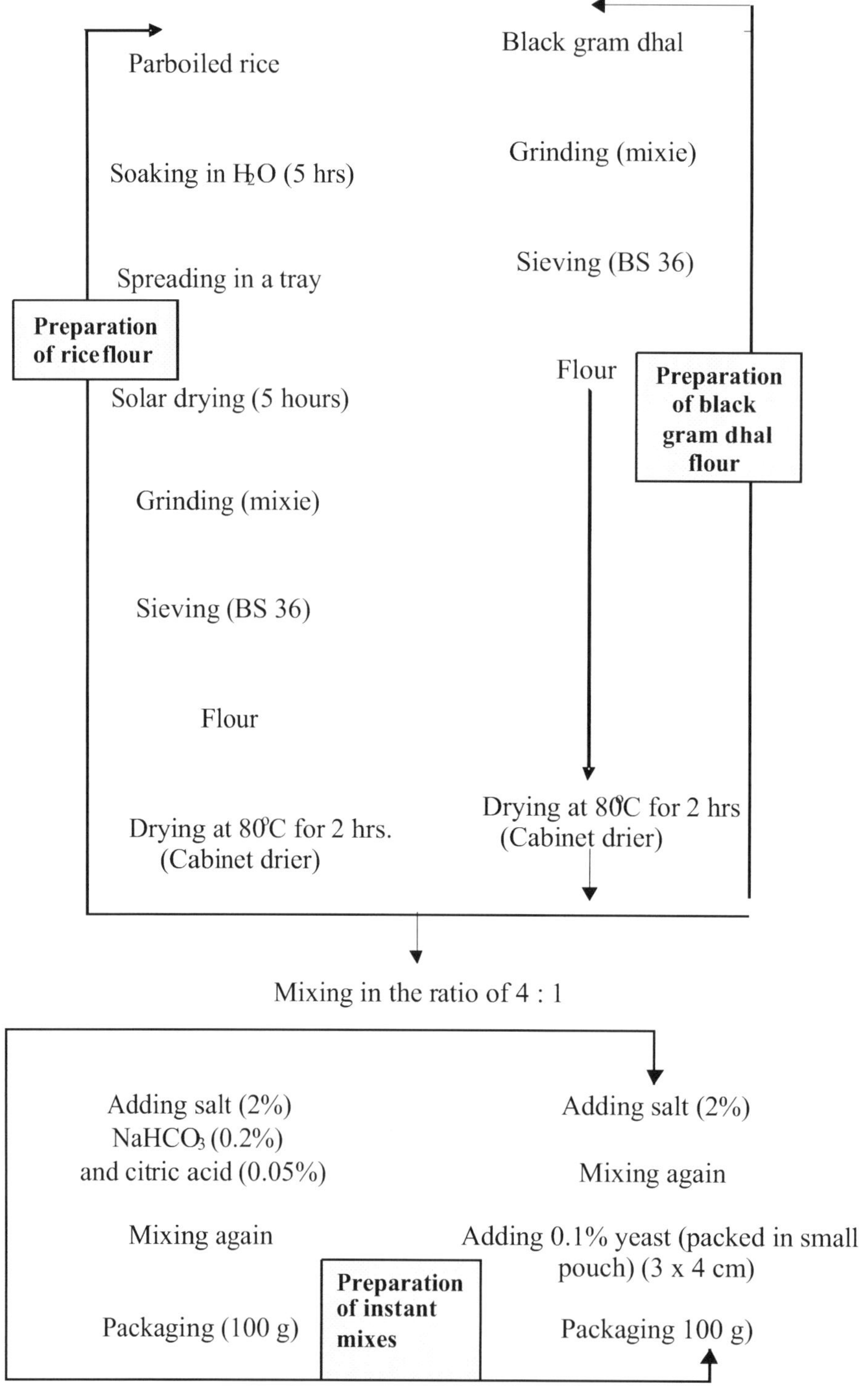

Figure 8: Flow diagram for the preparation of Instant idli mix

Preparation of Idli from Instant Mix

Batter Preparation

Instant Idli Mix Without Yeast

For one part of instant idli mix 0.5 part of fermented curd and 1.5 part of water were added and mixed well. The prepared batter was allowed to stand for 30 min at room temperature before preparing idlies.

Instant Idli Mix With Yeast

The yeast packet was opened and mixed with instant idli mix. For 100 g of instant mix 220 ml of water was added and mixed thoroughly. The batter was allowed to ferment for five hours at room temperature.

Preparation of Idli

Idli was prepared in an idli steamer. The prepared batter was poured into the plates and steamed for 10 min in the idli steamer. The prepared product was organoleptically evaluated.

Standardization and Preparation of Rava Dosa Ready Mix

Dosa is a fermented product used in India. This is usually prepared from a fermented batter of rice and pulse in the proportions ranging from 6:1 to 10:1. Both the ingredients are finely ground, unlike the idli batter, which contains the rice component in a coarse consistency. The dosa batter is very thin and dosa is baked on a hot pan. A number of varieties of dosa can be prepared in hotels and uses.

Among those masal dosai, onion dosai and rava dosai are highly popular irrespective of all the age groups.

Procedure

Preparation of Rice Flour

The parboiled rice – IR –20 was soaked in water for 5 hours. Then the water was drained completely and dried in solar drier for 5 hrs. The dried rice was ground in a mixie and sieved through BS 36 sieve.

Materials Required

Rice flour, maida, semolina (white ravai), salt, cumin, citric acid, dehydrated green chillies.

Table 14: Ingredients Combinations of rava dosa

Raw ingredients (g)	I	II	III	IV
Maida	100	100	80	80
Semolina	100	100	120	120
Rice flour	100	100	100	100
Cumin	5	5	5	5
Dried and powdered green chillies	4	4	4	4
Salt	6	6	6	6
Broken pepper	2	2	2	2
Citric acid	0.05% (5 mg)	Nil	0.05 %	Nil
$NaHCO_3$	0.2 % (20 mg)	Nil	0.2%	Nil
Yeast	Nil	0.1%	Nil	0.1%

Method

All the ingredients were mixed thoroughly and packed in an airtight container for future use.

Preparation of Rava Dosa

For one part (measure) of instant rava dosa mix two part (measure) of water was added and mixed well. The prepared batter was allowed to stand for 30 min at room temperature before preparing dosa. The dosa tava was heated and oil / ghee was smeared it well. One cup of batter was spread on the tava and baked on both the sides by using low flame.

Note

1. Finely chopped onions, green chillies, coriander leaves and curry leaves was also added to the batter for better taste.
2. Alternately, the dosa batter was also prepared in the same way by using ½ measure of curd and 1½ measures of water instead of two measures of water. The prepared product was organoleptically evaluated for standardization.

Standardization and Preparation of Instant Idiappam Mix

Idiappam is traditional food prepared from the combination of parboiled rice and raw rice. It is consumed as breakfast / dinner for the peoples especially South Indians. It is a steamed product and consumed as either in the form of sweet or savoury dish. (lime bath, tomato bath and curd bath).

Procedure

Preparation of Rice Flour

The parboiled rice – IR 20 and raw rice was soaked in water for 5 hours separately. Then the water was drained completely and dried in the solar drier for 5 hrs. The dried rice was ground in a mixie and sieved through BS 36 sieve.

Table 15: Materials required for rice flour

Ingredients (g)	I	II	III	IV
Parboiled rice flour	100	50	75	25
Raw rice flour	-	50	25	75
Salt	2	2	2	2

Method

Mix all the ingredients and stored, in the polythene bag.

Preparation of Idiappam

The idiappam mix (100 g) was steamed in a steaming unit for 5 minutes (idli cooker). To the steamed flour, required quantity of boiling water (90°C, 120 ml) was added and made into dough. The dough was extruded through hand noodle press, steamed in the idli cooker for 10 minutes. The prepared idiappam was organoleptically evaluated.

Pulses Based Convenience Foods

Preparation of Instant Adai Mix

Adai is a traditional food for the South Indians. It is prepared from the combination of cereals and pulses. It is a unfermented food. Legumes provide the protein supplement to the diet which primarily consists of carbohydrates. The complementary nutritive value of cereals and pulses suggest that the most practical means of eradicating the widespread protein calorie malnutrition in several areas of the world is to increase the supply of cereal pulse mixtures for human diets.

Method

Preparation of the Samples

Cleaning: The parboiled rice, raw rice, black gram dhal, red gram dhal, green gram dhal and bengal gram dhal were purchased from the local market, cleaned to remove immature, weeviled, broken and damaged grains, dried for two to three hours in sun, cooled and stored in the air tight containers till they were used.

Grinding of cereal and pulses: The cereals and pulses were ground in a mixie separately and sieved in BS 22 sieve. The ground flours were dried separately in a cabinet drier at 80°C for two hr, cooled and used for instant adai mix preparation. Mixing of ingredients **:** The ground cereals and pulses were mixed in the following proportions

Table 16: Raw ingredients used for Instant adai mixes

Raw ingredients (g)	Variation I	Variation II	VariationIII	VariationIV
Parboiled rice	200	-	300	120
Raw rice	200	100	300	120
Black gram	200	100	150	240
Green gram	200	250	150	240
Red gram	200	250	150	240
Bengal gram	200	500	150	240
Salt	20	20	20	20
Chilli powder	35	35	35	35
Asafoetida	3.0	3.0	3.0	3.0
Citric acid	1.0	1.0	1.0	1.0
Sodium bicarbonate	1.0	1.0	1.0	1.0
Total (g)	1260	1260	1260	1260

Method

The weighed cereal and pulse flours were mixed by using ladle continuously in a round bottom vessel by adding small quantum at a time to get uniform mixing. Salt, citric acid, sodium bicarbonate, chilli powder and asafoetida were added to it and mixed thoroughly.

Adai Preparation

Batter preparation: To 100 gm of instant adai mix, 200 ml of water was added and mixed well. The prepared batter was allowed to stand for 15 minutes at room temperature before preparing adai.

Other Ingredients Needed

Onion – 33 g, ginger – 1.3 g, garlic – 0.6 g, curry leaves – 3.3 g, oil – as required.

Note: The onion, ginger, garlic and curry leaves were cut into pieces and mixed with batter.

Preparation of Adai

The prepared batter was poured on the pre-heated tawa (Shallow fat frying pan) and spread uniformly with a ladder (thickness of 5 mm and diameter of 12 cm). The adai was cooked in the heated tawa by sprinkling oil on the surface for 2-3 ml. The prepared adai was light brawn in colour. Adai batter can also be prepared either with water and curd or water and tomato juice in the ratio of 1:1. The prepared adai was organoleptically evaluated.

Preparation of Instant Vada and Bajji Ready Mixes

Vada and bajji are the unfermented traditional snack items for the South Indians. The instant vada and bajji mixes are generally made from flours of bengal gram, black gram, wheat and rice. Frying not only aids in cooking, it also contributed to destruction of anti-nutritive factors, moisture evaporation and thereby increase shelf stability.

Method

Preparation of Instant Vada Mix

Grinding Pulses and Raw Rice

The bengal gram dhal, black gram dhal and raw rice were ground in a mixie separately. The ground bengal gram was sieved in a BS 14 sieve and the rest of the items were sieved in BS 36. Then the ground flours were dried separately in a cabinet drier at 80°C for two hours, cooled and stored in an airtight container till they were used.

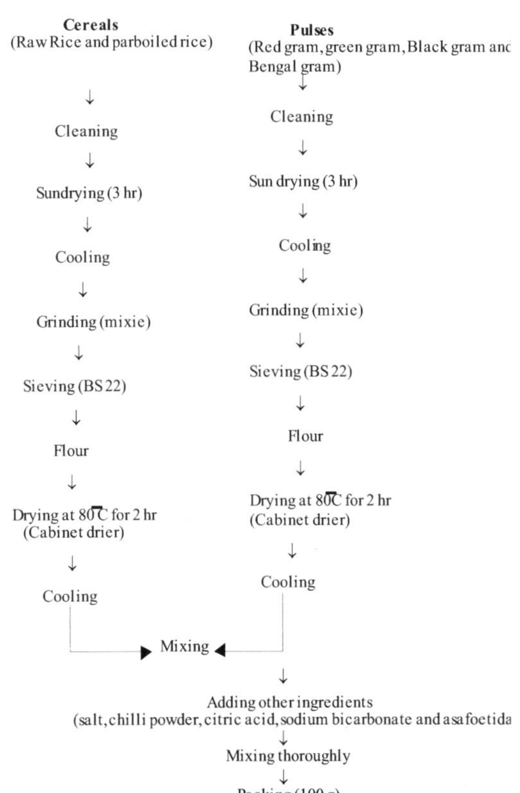

Figure 9: Flow diagram for preparation of instant adai mix

Preparation of Instant Vada Mix

Ingredients

Bengal gram dhal flour	:	85 g
Black gram dhal flour	:	5 g
Raw rice flour	:	5 g
Maida	:	2 g
Salt	:	2 g
Hydrogenated fat (dalda)	:	20 g
Citric acid	:	0.01 per cent
Sodium bicarbonate	:	0. 1 per cent

Method

1. The ground bengal gram dhal was mixed with maida and the flours of black gram dhal and rice.
2. The salt, citric acid and sodium bicarbonate were added to it and mixed well.
3. Melted cooled hydrogenated fat was mixed with the instant vada mix.
4. Another batch of vada mix was prepared without adding hydrogenated fat.

Other Ingredients Needed

Onion	:	15 g
Green chillies	:	15 g
Curry and coriander leaves	:	2 g
Aniseed	:	2 g
Ginger	:	5 g
Oil	:	150 ml
Water	:	65 ml/100 g

Method

1. The onion, green chilles, ginger, curry and coriander leaves were cut into pieces.
2. These were mixed with instant vada mix.
3. The dough was prepared by adding required quantity of water (65 ml/100g) and allowed to stand for 30 min.

4. The dough was made into a medium size balls and flattened into a round shape vada (dia 5.2 cm).
5. The oil (150 ml) was taken in a deep fat frying pan and heated to 150°C.
6. The vada was fried in the heated oil for 5 min. until light brown in colour.
7. The required quantum of dalda was added to the instant vada mix packed without dalda while preparing vada.

Preparation of Instant Bajji Mix

Table 17: Ingredients used for instant bajji mix

		Proportions (g)	
	I	II	III
Bengal gram flour	80	80	80
Black gram flour	5	-	-
Raw rice flour	5	10	20
Maida flour	10	10	-
Red chilli powder	10	7	7
Salt	3	3	3
Asafoetida	1	1	1
Sodium bicarbonate	0.1%	0.1%	0.1%

Method: All the ingredients were mixed thoroughly and sieved twice.

Other Ingredients Needed for Preparing Bajji

Big onions	:	50 g
Oil	:	150 ml
Water	:	120 ml / 100g

Method

1. Batter was prepared from instant bajji mix by adding water (120 ml/100 g).
2. Oil was taken in a frying pan and heated to 150°C.
3. Sliced onions were dipped in the batter and fried in the heated oil till it turns golden brown in colour.

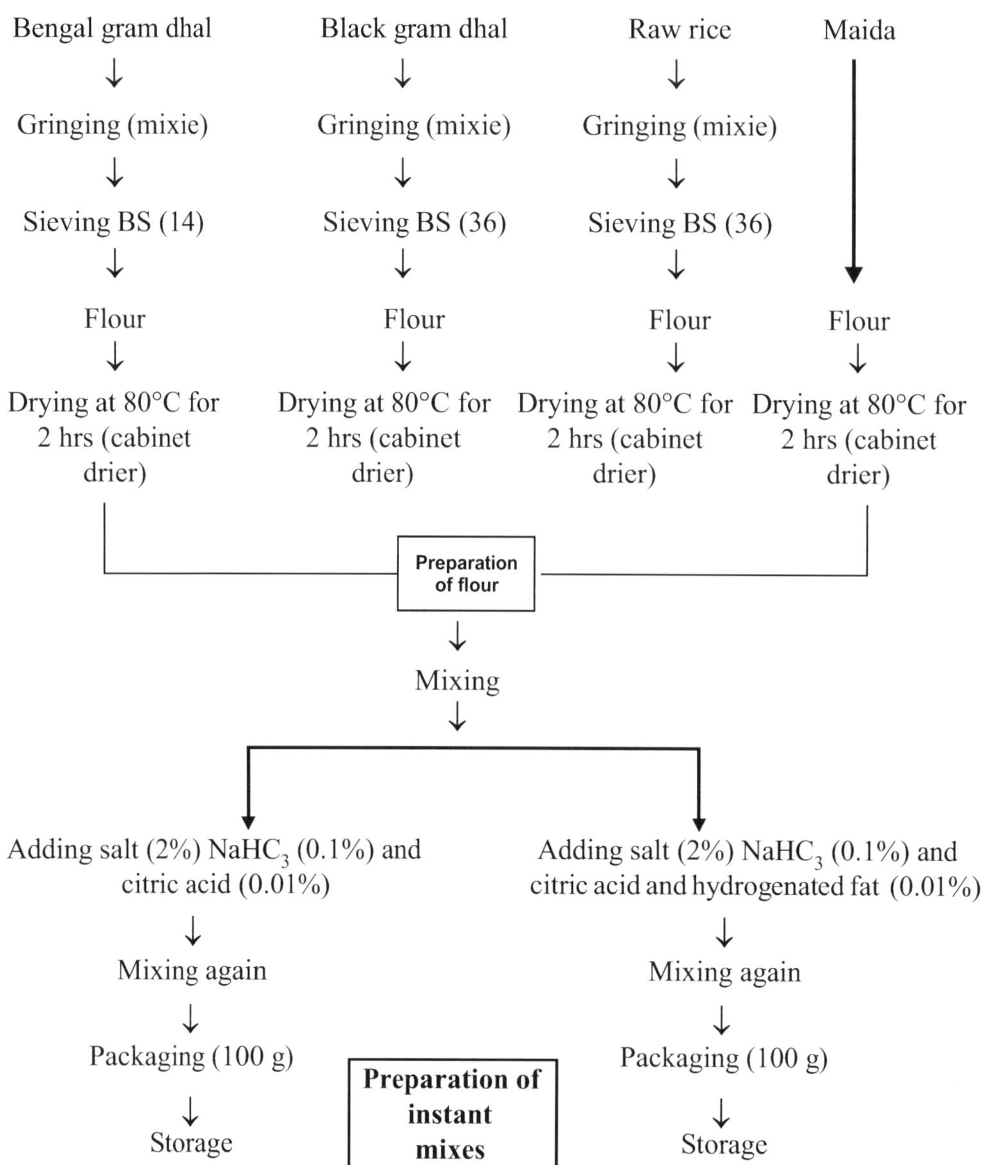

Figure 10: Flow chart for the preparation of instant vada mix

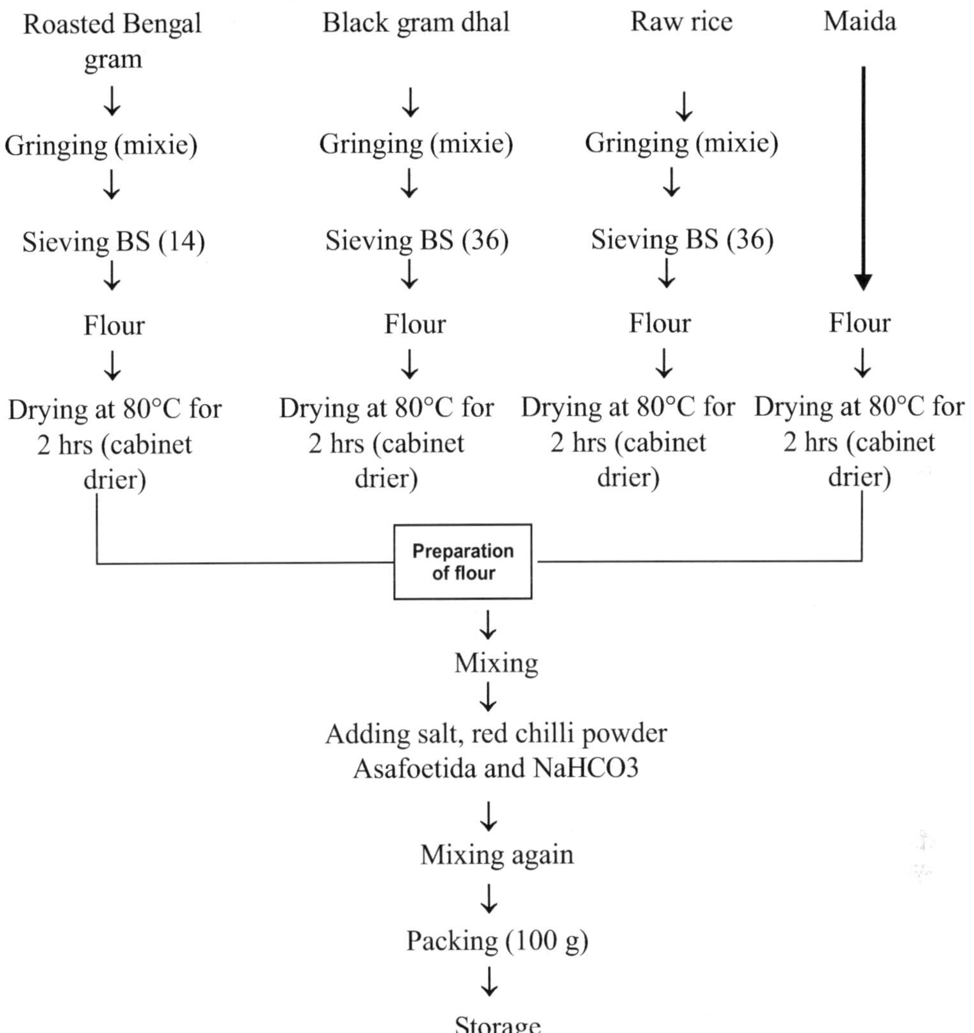

Figure 11: Flow chart for the preparation of instant bajji mix

Preparation of Pakoda and Murukku Instant Ready Mix

Pakoda and murukku are the snack items prepared by using pulse and cereal combinations. They are deep fat fried foods.

Method

Table 18: Ingredients used for Instant pakoda mix

Raw ingredients (g)	I	II
Bengal gram flour	100	100
Rice flour	25	25
Salt	3	3
Chilli powder	1	1
Sodium bicarbonate	0.1%	0.1%
Aniseed	1	1
Dalda	1	7.5

Method

All the ingredients were mixed thoroughly and sieved twice.

Preparation of Pakoda from Instant Mix

Big onions	:	2 nos.
Green chillies	:	2 nos.
Chopped curry and coriander leaves	:	2 gms
Oil	:	200 ml
Water	:	80 ml / 100 g

Method

1. A thick dough was prepared from instant pakoda mix by adding water.
2. Chopped onion, chopped curry leaves and coriander leaves were added to the batter and mixed well. The batter was kept for as such for 5 minutes.
3. Oil was taken in a frying pan and heated to 150°C.
4. The dough was fried in the heated oil till it turns golden brown in colour.

Preparation of Murukku Instant Mix

Method

Raw rice and black gram were ground in a grinder mill separately and sieved through 80 BS sieve and used for the preparation of murukku instant mix.

Table 19: Ingredients used for instant murukku mix

Ingredients	Variations					
	I	II	III	IV	V	VI
Rice flour (%)	100	98	96	94	92	90
Black gram flour (%)	-	2	4	6	8	10
Salt (g)	2	2	2	2	2	2
Water (ml)	86	86	88	90	92	92
White sesame seed	5	5	5	5	5	5

The crispy mix was prepared by mixing raw rice flour, black gram dhal flour, salt and sesame seed. The mix was passed through 80 BS sieve for uniform mixing.

Method for the Preparation of Murukku

The dough was prepared by the addition of hydrogenated fat –5 g and water and extruded in a hand extruder of local make with 4 mm diameter orifice and fried in the heated oil at 180°C for 3-5 minutes.

Soy Bean Based Value Added Products

Value-added soybean products—soy flour, textured soy protein, soy protein concentrates, and soy protein isolates, and soy milk replacer—offer improved nutrition, result in foods that have improved functional properties, and when consumed on a regular basis can be part of a program that will not only improve general health but prevent some chronic diseases including heart disease. Soy flour, soy protein concentrates, and soy protein isolates are all made from whole soybeans. Textured soy protein is made from either soy protein concentrate or soy flour.

Soy Flour

Soy flour is made from roasted soybeans ground into a fine powder. There are three kinds of soy flour available - natural or full-fat, which contains the natural oils found in the soybean and is 50% protein; defatted, which has the oils removed during processing and is an even more concentrated protein than full-fat soy flour; and lecithinated, which has lecithin added to it.

All soy flour gives a protein boost to recipes. Soy flour is gluten-free, so yeast-raised breads made with soy flour have a more dense texture. It is best used for fortification of other flours, including wheat, rice, and corn. Adding up to 10% of soy flour can greatly increase the protein and nutritional content of bread, chapattis, tortilla, and Arab pita bread, which are often the principle diet and at times the only staple that poor people get in some countries. Addition of soy to wheat or corn bread increases its shelf life and delays staling. Soy flour absorbs about twice as much water as its weight. It also retards oil absorption in fried foods.

Soy Protein Concentrates

Concentrates come from defatted soy flakes and contains at least 70 percent protein. It is a highly digestible source of amino acids and retains most of the beans' dietary fiber. The bland flavor of soy protein concentrates makes them an ideal source of protein in many foods including dietary products, pasta, tortillas, nutritional beverages, meat analogs, and processed meat products. Concentrates can be used to improve texture and mouth feels in foods while providing an excellent source of highly digestible protein. Even a small quantity can greatly improve the nutritional content of the food they are added to.

Functional properties of food can be improved by using soy protein concentrates. Water retention in breads can be increased and act as an emulsifier and bind fats. By adding soy protein concentrate to the outside of doughnuts, less excess fat is absorbed during frying. In meat products, they act as an emulsifier and bind water causing it to be juicier and loose less weight after cooking.

Soy Protein Isolates

When protein is removed from defatted flakes the result is soy protein isolate, the most highly refined soy protein. Containing 90 per cent protein, isolates possess the greatest amount of protein of all soy products. They are a highly digestible source of amino acids and because of the bland taste can be added to foods without jeopardizing flavor or characteristics.

Isolates are used to add juiciness, cohesiveness, and viscosity to a variety of meat, seafood, and poultry products. This is especially true when soy is used to enhance the flavor and nutritional quality of tough meat. It also does excellent job improving the sensory attributes of whole meat products. Roasts and hams that contain soy isolates will be juicer and more nutritional. In addition, isolates are commonly used in dairy type products such as beverages, frozen desserts and invitation cheeses. They can also be used as an ingredient to supplement or replace milk powder in a variety of uses.

Textured Soy Protein (TSP)

Textured soy protein (TSP) usually refers to products made from textured soy flour and textured soy protein concentrates. TSP can contain between 50% and 70% protein, depending on the starting material used. TSP is most commonly used as a meat extender or analog and can be added to a meal to increase its protein content. TSP has a texture similar to ground beef or other meat products and must be rehydrated with boiling water before use.

Textured Soy Flour

Textured soy flour is made by running defatted soy flour through an extrusion cooker, which allows for many different forms and sizes. It contains 50 percent protein as well as the dietary fiber and soluble carbohydrates from the soybean. When hydrated, it has a chewy texture. It is widely used as a meat extender. Often referred to simply as textured soy protein, textured soy flour is sold dried in granular and chunk style and has a bland flavor.

Textured Soy Protein Concentrates

Textured soy protein concentrates are made by extrusion and are found in many different forms and sizes. Textured soy protein concentrates contain 70 percent protein as well as the dietary fiber from the soybean. When hydrated, they have a chewy texture and contribute to the texture of meat products.

Soy Milk Replacer

Soy Milk Replacer has 26% protein and 39% carbohydrates. It has exactly the same nutritional properties as dairy milk. The starting material is soy concentrate, and several nutrients and other ingredients are added to give it the nutritional properties of dairy milk. It provides higher benefits than dairy milk, because it is lactose free and cholesterol free. It is a different product than soy milk and has a different nutritional profile. Soy milk replacer has to be reconstituted with water at a ratio of 1:8. It has packaging and storage requirements similar to dehydrated cow's milk.

Soy-Based Foods

Infant Formulas

Soy-based infant formulas are similar to other infant formulas except that a soy protein isolate powder is used as a base, instead of cow's milk. Carbohydrates and fats are added to achieve a fluid similar to breast milk.

Lecithin

Extracted from soybean oil, lecithin is used in food manufacturing as an emulsifier in products high in fats and oils. It also promotes stabilization, antioxidation, crystallization and spattering control. Powdered lecithins can be found in natural and health food stores.

Meat Alternatives (Meat Analogs)

Meat alternatives made from soybeans contain soy protein or tofu and other ingredients mixed together to simulate various kinds of meat. These meat alternatives are sold as frozen, canned or dried foods. Usually, they can be used the same way as the foods they replace. With so many different meat alternatives available to consumers, the nutritional value of these foods varies considerably. Generally, they are lower in fat, but read the label to be certain. Meat alternatives made from soybeans are excellent sources of protein, iron and B vitamins.

Nondairy Soy Frozen Dessert

Nondairy frozen desserts are made from soymilk or soy yogurt. Soy ice cream is one of the most popular desserts made from soybeans and can be found in natural food stores.

Soy Cheese

Soy cheese is made from soymilk. Its creamy texture makes it an easy substitute for sour cream or cream cheese and can be found in variety of flavors in natural foods stores. Products made with soy cheese include soy pizza.

Soy Fiber (Okara, Soy Bran, Soy Isolate Fiber)

There are three basic types of soy fiber: Okara, soy bran and soy isolate fiber. All of these products are high-quality, inexpensive sources of dietary fiber. Okara is a pulp fiber by-product of soymilk. It has less protein than whole soybeans, but the protein remaining is of high quality. Okara tastes similar to coconut and can be baked or added as fiber to granola and cookies. Okara also has been made into sausage. Look for okara in natural food stores.

Soy bran is made from hulls (the outer covering of the soybean), which is removed during initial processing. The hulls contain a fibrous material which can be extracted and then refined for use as a food ingredient. Soy isolate fiber, also known as structured protein fiber (SPF), is soy protein isolate in a fibrous form.

Soynut Butter

Made from roasted, whole soynut which are then crushed and blended with soyoil and other ingredients, soynut butter has a slightly nutty taste, significantly less fat than peanut butter and provides many other nutritional benefits as well.

Soy Yogurt

Soy yogurt is made from soymilk. Its creamy texture makes it an easy substitute for sour cream or cream cheese. Soy yogurt can be found in variety of flavors in natural foods stores.

Whipped Toppings, Soy-Based

Soy-based whipped toppings are similar to other nondairy whipped toppings, except that hydrogenated soyoil is used instead of other vegetable oils.

Traditional Soy Foods

"Traditional" soy foods are developed from whole soybeans, the edible seed of the soybean plant. They are high in protein and contain beneficial phytochemicals, such as isoflavones. Whole soybeans can be cooked and used in sauces, stews, and soups. Whole soybeans that have been soaked can be roasted for snacks.

Miso

Miso is a rich, salty condiment that characterizes the essence of Japanese cooking. The Japanese make miso soup and use it to flavor sauces, dressings, marinades, and pâtés. A smooth paste, miso is made from soybeans and a grain such as rice, plus salt and a mold culture, and then aged in cedar vats for one to three years. Miso comes in many different colors, textures and grades.

Natto

Natto is made of fermented, cooked whole soybeans. Because the fermentation process breaks down the beans' complex proteins, natto is more easily digested than whole soybeans. It has a sticky, viscous coating with a cheesy texture. Natto is traditionally served as a topping for rice, in miso soups, and is used with vegetables.

Okara

Okara is a pulp fiber by-product of soymilk or tofu. It has less protein than whole soybeans, but is a nutritional powerhouse, containing soluble and non-soluble fiber, protein, calcium and other minerals. Okara tastes similar to coconut and can be baked or added as fiber to granola and cookies. The traditional Japanese way of eating okara is to flavor it by stir-frying with dark sesame oil and soy sauce, then mix it together with vegetables or put it into a soup.

Soy Milk

Soybeans soaked, ground fine, and strained produce a fluid called soymilk. Soy milk is a good source of protein, thiamine, iron, phosphorous, copper, potassium and magnesium. It contains little sodium. Some brands are fortified with important vitamins and minerals such as calcium, Vitamin D, and Vitamin B-12. Soy milk also is low in saturated fat and is cholesterol-free.

Soy Nuts

Roasted soynuts are whole soybeans that have been soaked in water and then baked until browned. High in protein and isoflavones, soynuts are similar in texture and flavor to peanuts. Soy nuts are primarily used as a snack item but their crunchy texture and nutty flavor also make an interesting addition to salads and grain dishes.

Soy Sauce

Soy sauce is a dark brown liquid made from soybeans that has undergone a fermentation process. Soy sauces have a salty taste, but are lower in sodium than traditional table salt. Specific types of soy sauce are shoyu, tamari, and teriyaki. Shoyu is a blend of soybeans and wheat. Tamari is made only from soybeans and is a by-product of making miso. Teriyaki sauce can be thicker than other types of soy sauce and includes other ingredients such as sugar, vinegar, and spices.

Soy Sprouts

Soy sprouts (also called soybean sprouts), are an excellent source of nutrition, packed with protein and vitamin C. They can be sprouted in the same manner as other beans and seeds. Soy sprouts must be cooked quickly at low heat so they don't get mushy. They can also be used raw in salads or soups, or in stir-fried, sautéed, or baked dishes.

Tempeh

Tempeh is a made by a natural culturing and controlled fermentation process that binds soybeans into a cake form. It is sometimes mixed with another grain such as rice or millet. The fermentation process and its retention of the whole bean give tempeh a higher content of protein, dietary fiber and vitamins compared to tofu, as well as firmer texture and stronger flavor. Tempeh can be marinated and grilled and added to soups, casseroles, and chili.

Tofu

Tofu, also known as soybean curd, is a soft cheese-like food made by curdling fresh hot soymilk with a coagulant. Tofu is a bland product that easily absorbs the flavors of other ingredients with which it is cooked. Tofu is rich in both high-quality protein and B vitamins and low in sodium. Firm tofu is dense and solid and can be cubed and served in soups, stir fried, or grilled. Firm tofu is higher in protein, fat, and calcium than other forms of tofu. Soft tofu is good for recipes that call for blended tofu. Silken tofu is a creamy product and can be used as a replacement for sour cream in many dip recipes.

Yuba (Also Known as Tau-kee)

Yuba, also known as Tau-kee, is a brittle soy wafer made by lifting and drying the thin layer formed on the surface of cooling hot soymilk. It has a high-protein content and is commonly sold fresh, half-dried, and dried. There are various names used for the soy wafer depending on the "thinness".

Quality Evaluation of Pulse Based Instant Mixes

The qualities of the pulse based instant mixes are judged both subjective and objective methods. The physical methods such as bulk density and oil absorption can be commonly used. The chemical methods such as moisture, fat, free fatty acid and peroxide value are estimated before and during storage. Organoleptic evaluation is done after preparing all the products from instant mixes. Microbial load is carried out before and during storage of instant mixes.

Procedure

1. **Bulk density:** The prepared product was weighed accurately. The volume of the weighed product was measured by mustard replacement method. The weight by volume was expressed as bulk density.
2. **Moisture:** The moisture content of instant mix- samples were estimated by the air oven drying method. The sample was dried at 110ºC. The drying was continued till a constant reading was obtained. The moisture content was expressed as percentage.
3. **Free fatty acid:** A known weight (5 to 10 gm) of the sample prepared product was taken into a clean dry conical flask. To this 25 ml of 95% alcohol was added. A few drops of phenolphthalein indicator was added and titrated against 0.1 N KOH. The end point was the appearance of pale pink colour which persists for 30 seconds.

Calculation

$$\text{Free fatty acid content} \quad \frac{\text{Titre value} \quad \text{NKOH} \quad 56.1}{\text{Weight of the sample(g)}}$$

1 ml of N/10 KOH = 0.0282 gm of oleic acid

References

Aykrod, W.R. and J.Doughty, Legumes in Human Nutrition, FAO, Rome, 1973.

Burrington, K.J. Keeping the crunch in breakfast cereals, Food Prod. Design, P.63. June 2001.

Cheng, L.M. 1992. Food Machinery, For the Production of Cereal Foods, Snack foods and Confectionary. Ellis Horwood, New York.

Fennema, O.R. 1996. Food chemistry. Third edition. Marcel Dekker Inc. NY. pp. 195-534.

Indian Council of Agricultural Research, Wheat Research in India, New Delhi, 1966.

Indian Council of Agricultural Research, Jowar, New Delhi, 1980.

ISI (1969) Specifications for processed cereal weaning foods. No. 1656. New Delhi: Indian Standards Institution.

Kacharoo, P., Pulses crops of India, ICAR, New Delhi, 1970

Kent, N.L., Technology of Cereals, Pergamon Press, New York, 1978.

Kuntz, L.A. Breakfast Cereals Grow Up. Food Prod. Design. -.84, August, 2000

Malleshi NG (1993) Processing of coarse cereals for food and industrial uses. In: Food Technology Highlights Souvenir. 3rd International Food Convention. Mysore: Assoc. of Food Scientists and Technologists, p. 58.

Marz, S.A., Cereal Sciences, The AVI Publishing Co. Inc., Connecticut, 1969

Marz, S.A., Bakery Technology and Engineering The AVI Publishing Co. Inc., Connecticut, 1978.

Marz, S.A., and T.D.Martz Cookie and Craker Technology,, The AVI Publishing Co. Inc., Connecticut, 1972.

Murthy, R.B., Breeding Procedures in Pearl Millet, ICAR, New Delhi, 1977

Pal. B.P., Wheat, ICAR, New Delhi. 1966.

Rokey, G.J. RTE Breakfast Cereal Flake Extrusion, Cereal Foods. Word.40: 422, 1995

Sultan, W.J., Modern Pastry Cheff, The AVI Publishing Co.Inc., Connecticut, 1977

Vessa, J.A. Processing characteristics of kneading single screw vs corotating twin screw cereal extrusion, Cereal Foods World 35: 1162, 1990

Watson, E.L., Harper, J.C., and Harper, J.C.1988. Elements of Food Engineering. Chapman & Hall, London, New York

Whiteley, P.P., Biscuit Manufacture, Elsevier Publishing Co. Ltd., London, 1971.

Williams, A, (ed), Bread Making, Hutchinson, Benhan, London, 1975

Chapter - 2

Traditional Foods

Importance of Traditional Foods

Traditional convenience food, or tertiary processed food, is commercially prepared food designed for ease of consumption. Products designated as convenience foods are often prepared food stuffs that can be sold as hot, ready-to-eat dishes; as room-temperature, shelf-stable products; or as refrigerated or frozen products that require minimal preparation (typically just heating).

These products are often sold in portion controlled, single serve packaging designed for portability for "on-the-go" eating. Convenience food can include products such as candy; beverages such as soft drinks, juices and milk; fast food; nuts, fruits and vegetables in fresh or preserved states; processed meats and cheeses; and canned products such as soups and pasta dishes.

Traditional Foods

Preparation of Vadagam and Appalam

Vadagam is a traditional product mostly prepared from rice flour and also sago. It is an deep fat fried product and consumed as a side dish in the meals. Papad also known as appalam or papadam in South India, is essentially a thin wafer like product, usually circular in shape rolled from a dough made out of pulse, cereals, edible vegetable oil, alkaline and mucilaginous additives. It is normally consumed in toasted or fried form. In India, papad has been associated and deeply entwined with social customs/ rituals and has been developed into culinary art.

Preparation of Vadagam

Materials Required

Raw rice flour, green chillies, cumin seeds and salt.

Raw Rice flour	-	100 gm
Green chillies	-	1.5 gm
Cumin seeds	-	4.0 gm
Salt	-	2.0 gm

Method

The ingredients were mixed and 250 ml of water was added and cooked for 10 minutes until a thick extrudable paste was formed. The hot paste was extruded in hard extruder (dia 5 mm) and dried in sun for about five hours. The dried vadagam sample was packed in a polyethylene bags and sealed.

Frying of Vadagam

Dried rice vadagam was fried in the refined oil at 180°C for 40 seconds. The vadagam was organoleptically evaluated.

Preparation of Papad

Materials Needed

Black gram flour	-	100g
Water	-	50 ml
Salt	-	7 gm
Sodium bicarbonate (or) Sodium carbonate	-	1 gm
Asafoetida	-	0.2 gm
Refined oil	-	2-3 gm

Method

All the ingredients except water were mixed well. The dough was prepared after adding water. The prepared dough was kneaded well and divided into balls of 15 gm each and pressed to 0.8 cm thickness with 8-9 cm diameter papad. The papad was soaked overnight, conditioned and dried under the sun for 10 minutes to reduce the moisture content to $\pm$ 14.0 per cent. It was packed in polyethylene pouches. The steps involved in the processing of papad are given below:

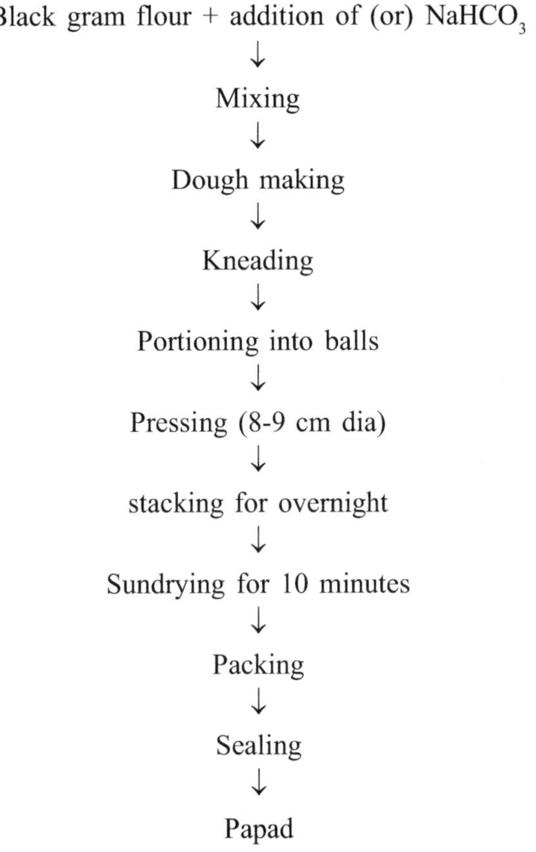

Fig 12: Flow diagram of Processing of papad

Frying of Papad

Dried papad was fried in the refined oil at 195°C for 5 to 7 seconds. The papad was organoleptically evaluated.

Frying Characteristics of the Appalam and Vadagam

Frying characteristics of the vadagam and papads were studied by deep fat frying in vegetable oil at 190°C. The volume expansion was calculated by measuring the length, breadth and diameter expansion before and after frying. The blister formation in papads was also noted. The fat absorption found out by weighing the sample before and after frying.

Preparation of Khakra

Khakra is an Indian flat bread, like chapathi and is a popular traditional food in Western India, particularly Gujarat. Atta (whole wheat flour) is kneaded with water into chapathi dough consistency and rolled (flattened) to 1 to 1.5 mm thickness and 17 to 25 cm diameter. These are toasted on a hot pan (tawa), with frequent pressing of the khakra for uniform contact heating. Some of the commercially available khakra are masala khakra which is prepared by adding good amount of spices, bhoji khakra is prepared by incorporating green leafy vegetables, ghee khakra is prepared by adding ghee, sweet and plain khakra. The crisp texture and long storage life of khakra are two characteristics which make it a convenience food constituting one of the major items in lunch packs, breakfast foods, anytime snack or in food packs during travel. They are served accompanied by Sabji (cooked vegetables). Pickle and sweet or hot chutney or can be even dipped in tea or coffee and eaten. The price of khakra is also comparable to that of chapatis.

Preparation

Ingredients Needed

Wheat flour	-	100 g
Salt	-	3 g
Water	-	About 75 ml

Method

1. To 100 gm of wheat flour, 3 gm of salt and sufficient amount of water was added, to prepared a dough.
2. Kneaded for 10 minutes to obtain a dough.
3. The dough was portioned into 15 gm balls. Each ball of dough was rolled into then chapati using a corrugated aluminium roller, which would prevented puffing of khakra during toasting. The khakra was rolled to a thickness of 1.25 mm and 15 cm diameter.
4. Khakra were toasted on tawa. The khakra was pressed and turned frequently with a cloth bundle (a thick cloth was folded into a 10 cm square of thickness 5-6 cm. The folded material was placed in another piece of cloth and made into a bundle, so that it was handy to use. This cloth bundle was used to press the khakra while toasting. The pressing process ensured even distribution of heat all over the khakra). The khakra was cooked for 5 minutes with temperature of 160°C. The prepared khakra was organoleptically evaluated.

Preparation of Ready Mixes (Masala and Soup)

A. Preparation of Sambar Powder

Ingredients: Coriander seeds – 1 kg, Red chillies – 1 kg, Pepper – 100 gm, Turmeric powder – 10 gm, Red gram dhal – 200 gm, Bengal gram dhal –100 gm, Fenugreek – 100 gm, Cumin seeds – 100 gm Asafoetida – 20 gm, Curry leaves – 100 gm, oil – for frying.

Method

1. Dry the red chillies in the cabinet drier at 60°C for 2 hrs or in the sun for 5 hours.
2. Fry the asafetida, fenugreek and cumin in oil separately.
3. Fry the remaining ingredients separately till it gets flavour without adding oil.
4. Mix all the ingredients and powder it in the mill.
5. Cool it and packed either polythene bags or air tight glass containers.

B. Preparation of Rasam Powder

Ingredients: Coriander seeds – 1 kg, Red chillies – 600 gm, red gram dhal – 400 gm, pepper – 400 gm, turmeric powder – 20 gm, cumin – 200 gm, fenugreek – 40 gm, Asafoetida – 40 g curry leaves – 100 gm, oil – for frying.

Method

Fry the red chillies in oil till it become light brown in colour. Fry the other ingredients viz., coriander seeds, dhal, pepper, cumin without adding oil till it becomes golden brown. Grind separately in each ingredient and mix it finally. Pack the prepared material in polythene bags or an air light glass container/ plastic containers.

C. Preparation of Chicken Mix

Ingredients: Red chillies – 1 kg, coriander seeds – 500 g, pepper – 100 g, salt – 200 gm, cumin – 100 gm, dried and powdered ginger – 10 gm, cloves – 5.0 gm, cinnamon – 10.0 gm, cardamom – 5.0 gm, turmeric powder – 10 gm, dried and powdered garlic – 30 gm, dried and powdered onion – 100 gm refined groundnut oil – 50 ml for frying the ingredients.

Method: Fry the red chillies in oil till it becomes light brown in colour. Fry the other ingredients viz., coriander seeds, pepper, cumin, cloves, cinnamon and cardamom separately. All the ingredients are mixed and made into powder in a machine (Huller). The powder is cooled and packed in polythene bags or an air tight glass bottles / plastic containers.

D. Preparation of Mutton Mix

Ingredients: Red chillies – 1 kg, coriander seeds – 750 gm, pepper – 50 gm, salt – 200 gm, cumin – 100 gm, dried and powdered ginger – 10 gm, cloves – 5.0 gm, cinnamon – 5.0 gm cardamom – 5.0 gm. turmeric powder – 10 gm, dried and powdered garlic – 20 gm, dried and powdered onion – 200 gm. Refined groundnut oil – 50 ml for frying the ingredients.

Method: Fry the red chillies in oil till it becomes light brown in colour. Fry the other ingredients viz., coriander seeds, pepper, cumin, cloves, cinnamon and cardamom separately. All the ingredients are mixed and made into powder in a machine (Huller). The powder is cooled and packed in polythene bags or an air tight glass bottles / plastic containers.

E. Preparation of Fish Gravy Mix

Ingredients: Red chillies – 1 kg, coriander seeds –300 gm, salt – 150 gm, cumin – 100 gm, dried and powdered ginger – 20 gm, dried and powdered garlic – 20 gm.

Method: Dry all the ingredients in sun for 5 hrs (or) cabinet drier at 80°C for 2 hrs. Mix all the ingredients and made into powder by using machine (Huller). Cool the powder and packed in polythene bags / air tight glass or plastic containers.

F. Preparation of Vegetable Masala Mix

Ingredients: Red gram – 50 gm, coriander seeds – 125 gm, cumin – 8 gm, cinnamon – 3 gm, cloves – 1 gm, pepper – 7.5 gm poppy seed – 200 gm, ani seeds – 2.0 gm, turmeric – 5 gm, red chillies – 2 gm, asafoetida – 2 gm, salt for taste.

Method: Fry all the above mentioned ingredients separately grind into a powder consistency using mixie (or) machine. Sieve the powder and pack it.

G. Preparation of Idli Podi

Ingredients: Redgram dhal – 75 gm, black gram dhal – 25 gm, dried chillies – 4 Nos, white sesame seed – 10 gm, asafoetida powder – 3 gm, curry leaves – 10 gm, salt – 5.0 gm, garlic powder – 5 gm, refined groundnut oil – for frying.

Method: Fry red gram dhal and black gram dhal in oil till the flavour comes out. Fry dried chillies, white sesame seed and curry leaves separately. Powder all the ingredients separately and cool it. Mix all the powdered ingredients and pack it either polythene bags or air tight plastic containers.

H. Preparation of Paruppupodi

Ingredients: Red gram dhal – 500 gm, bengal gram – 400 gm, dried chillies – 15 nos, black gram dhal – 50 gm, cumin – 10 gm, pepper – 10 gm, curry leaves – 10 gm, asafoetida – 5 gm, salt – 150 gm.

Method: Fry all the ingredients separately and made into powder by using machine (Huller). Mix all the ingredients, sieve it and pack it either polythene bags or air tight plastic containers.

Preparation of Instant Soup Mixes

A. Mushroom Soup Mix

Preparation of mushroom powder: Fresh oyster mushroom was washed in clean, cool water to remove the dirt. The cleaned mushroom was blanched in hot water for 3 minutes. Then it was sun dried for 8 hours (or) cabinet drier for 4 hours at 50-60°C. After drying the mushroom was powdered by using mixie. The powder was sieved.

B. Preparation of Mushroom Soup Mix

Ingredients: Mushroom powder – 50 gm, ragi flour – 20 gm, roasted bengal gram flour – 10 gm, pepper and cloves – 5 gm, cashewnut – 5 gm, cardamom – 1 gm, milk powder – 20 gm and salt – 10 gm.

Method

1. Roasted bengal gram, pepper, cashewnuts and cloves was powdered separately.
2. Mushroom powder was mixed with ragi flour, milk powder, pepper powder, cashewnut, cardamom and salt.
3. Soup mix was dried in cabinet drier at 80°C for 30 min.
4. The cooled mix was packed in polythene bags for storage.

C. Tomato Soup Mix

Preparation of Tomato Powder

Fully ripe and firm tomatoes were washed well in running tap water. Then it was cut into small pieces and dried in the cabinet drier at 80°C for 10 hours. The dehydrated pieces were then ground into powder in a mixie.

D. Preparation of Tomato Soup Mix

Ingredients: Tomato powder – 5.0 gm, onion powder – 0.5 g corn flour – 2.0 gm, cumin powder – 0.5 gm, pepper powder – 0.3 gm, salt – 1.5 gm, Aginomotto -0.5 gm.

Method: All the ingredients were mixed thoroughly and packed in polythene bags.

Preparation of onion powder: Disease free big onions were selected and the skin was peeled off. Then it was washed in the running tap water and cut into small pieces. It was dried in the cabinet drier at 60°C for 7 hrs. The dehydrated flakes were ground into powder and packed in polythene bags.

E. Vegetable Soup Mix

Ingredients: Onions, carrot, beans, cauliflower, cabbage, tomato and spinach.

Preparation of Vegetable Powders

All the selected vegetables were washed well in running tap water. Then it was cut into small pieces and steam blanched for 3.0 – 5.0 minutes except onion and tomatoes. All the vegetables except onion (60°C and 7 hrs) were dried in the cabinet drier at 80°C for 10 hrs separately. The dehydrated pieces were then ground into powder in a mixie.

Preparation of Mixed Vegetable Soup Mix

Vegetable powders – 50 gm (carrot: beans : cauliflower : cabbage: tomato : spinach 1:1:1:1:1:1)

Onion powder	:	5 gm
Corn flour	:	20 gm
Cumin	:	5 gm
Pepper	:	3 gm
Salt	:	15 gm
Agenomatto	:	5 gm

Method: All the ingredients were mixed thoroughly and packed in polythene bags.

F. Corn Soup Mix

Ingredients: Baby corn, onions and tomato.

All the vegetables were washed well in running tap water. Then it was cut into small pieces and steam blanched for 3.0 – 5.0 min. except onion and tomatoes. Then it was dried in the cabinet drier at 80°C for 10 hours except onion (60°C and 7 hrs). The dehydrated pieces were then ground into powder in a mixie.

Ingredients

Baby corn powder	:	25 gm
Tomato powder	:	25 gm
Onion powder	:	5 gm
Corn flour	:	20 gm
Cumin powder	:	5.0 gm
Pepper powder	:	3.0 gm

Salt	:	1.5 gm
Agenomotto	:	0.5 gm

Method: All the ingredients were mixed thoroughly and packed in polythene bags.

Quality Evaluation of Masala and Soup Ready Mixes

I. Masala Ready Mixes

Ingredients

Chicken-1.0 kg, chopped onion – 200 gm, chopped tomato – 200 gm, oil – 150 ml, coconut paste – 150 gm, salt to taste, chicken masala mix – 50 gm, curry leaves – for garnishing.

Method

1. Heat the oil and add mustard seed.
2. Season it with chopped onion till it becomes light brown.
3. Add chopped tomato and cook for 5 minutes.
4. Add chicken, chicken masala, salt and cook by adding sufficient amount of water for 30 minutes.
5. Add coconut paste and cook for again 3 minutes and garnish it with chopped curry leaves.

The above prepared chicken gravy was organoleptically evaluated by using ten untrained judges and its quality was evaluated.

II. Soup Mixes

Soup is a liquid food made by cooking vegetables, meat etc. together in a stock or water. It normally serves as a pleasant opening to a large meal particularly in cold weather and forms a good vehicle for vegetable, meat, milk, cheese etc. Soups provide nutritive values (appetizer), palatability and easy digestibility at low cost. Generally soups are prepared with fresh vegetables which require peeling, cutting, boiling, straining and seasoning. This makes the task tedious and time consuming. Therefore, preparation of instant soup mixes can be a boon against the time consuming culinary work. Simple techniques are involved in their preparation. A large number of soup mixes of different varieties are available in the world market.

A. Mushroom Soup

Ingredients: Mushroom soup mix – 10 gm, onion – 5 gm, green chillies – 2 nos, tomato – 10 gm curry leaves – 2 gm, coriander leaves – 5 gm, butter – 2 gm, water – 250 ml, salt – to taste.

Method

1. Tomato, chillies and coriander leaves were cut into small pieces.
2. Heat the oil and sauté all the vegetables except coriander leaves.
3. Add 250 ml of water and boil it.
4. After boiling, add mushroom soup mix and boil for another five minutes.
5. Then add butter and garnish with coriander leaves at the time of serving.

B. Tomato Soup

Ten gram of the prepared tomato soup mix was added to 150 ml of boiled water and stirred thoroughly.

C. Mixed Vegetable Soup

Ten gram of the prepared mixed vegetable soup mix was added to 150 ml of boiled water and stirred thoroughly.

D. Corn Soup

Ten gram of the prepared corn soup mix was added to 150 ml of boiled water and stirred thoroughly.

All the prepared soups were organoleptically evaluated by using 10 untrained judges.

III. Traditional Sweet

Sweet product	**: Ingredients needed**
Sample	: 100 gm
Coconut scraping	: 40 gm
Sugar	: 20 gm
Ghee	: 10 gm

Method: The sample was steamed in the idli steaming vessel for 15 minutes. Before serving, ghee, coconut scraping and sugar were mixed with the steamed noodle.

Payasam	**: Ingredients needed**
Sample	: 100 gm

Milk	:	100 ml
Sugar	:	50 ml
Cashew nut	:	10 gm
Raisin	:	5 gm
Powdered cardamom	:	2 gm

Method: Milk was boiled and sugar was added followed by slightly roasted sample was added to it and cooked well. Cashew and raisin (ghee) and powdered cardamom was added. Continued cooking for 5 minutes and served hot.

Convenience Sweet Ready Mix

A. Kheer mix

Traditional foods occupy an important place in the Indian dietary. Wheat and rice form the basic ingredients for several regional traditional foods. Kheer (or) payasam is one such item, which is highly popular with all households.

Raw rice was ground in plate grinder and sieved. The rice grits retained it 30 BS sieve and grits passed through 22 BS were used for preparing instant kheer mix. The rice grits were dried in Kilburn direr for 2 hours at 80°C and cooled.

Ingredients

Raw rice flour	:	100 gm
Sugar	:	80 gm
Milk powder (skimmed)	:	60 gm
Cardamom	:	2 gm
Salt	:	2 gm

Method: Instant kheer mix with the above said proportion of various ingredients were mixed and packed in polythene bags.

B. Carrot kheer mix

Fresh good quality carrots procured from the local market were washed thoroughly with water, trimmed and peeled in an abrasive peeler and cut into slices (2.5 cm length and 4 mm thick) using a dicer. Carrot slices were blanched in boiling water containing 0.12%, KMS for 3 min and soaked in 10%, sugar solution for 2 hrs. The slices were drained, dried in a cabinet drier at 60-70°C.

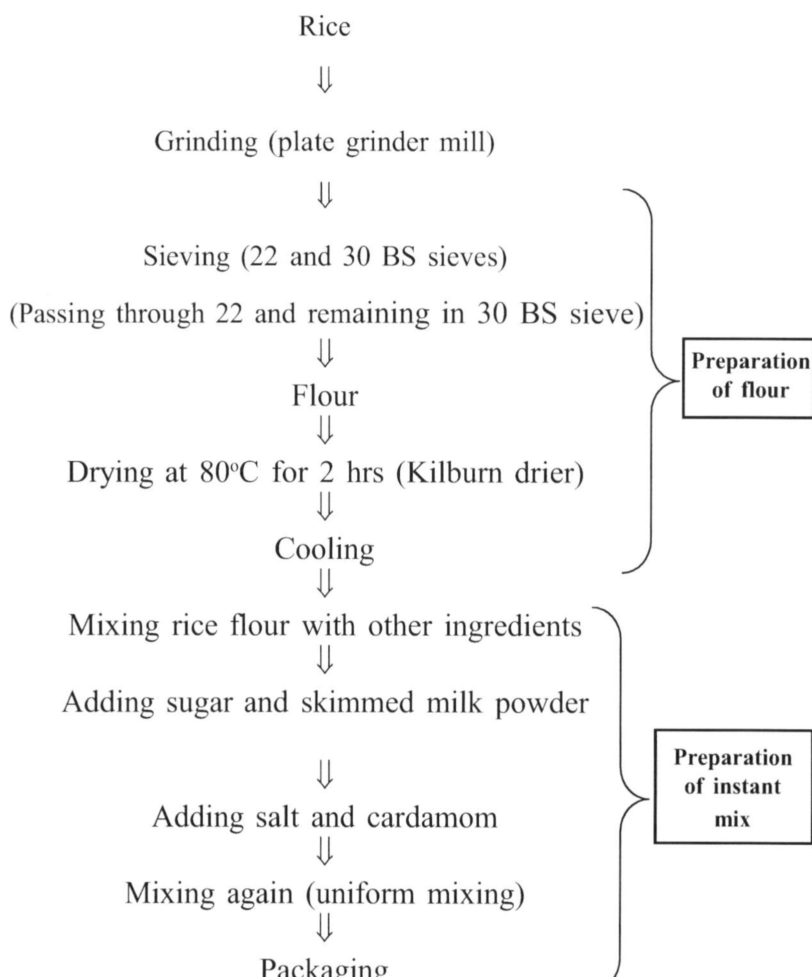

Figure 13: Flow chart for the preparation of Kheer mix

Ingredients

Dried carrot	: 10.16 gm
Sugar	: 47.43 gm
Skimmed milk powder	: 33.86 gm

Fried cashew nut	: 6.10 gm
Corn flour	: 2.03 gm
Cardamom powder	: 0.42 gm
Total	**: 100.0 gm**

The other ingredients of the mix like sugar, skim milk powder, cashewnut, cardamom and corn flour were also incorporated in the mix. Sugar and cardamom were powdered, cashewnut halves were fried to brown colour and corn flour was roasted before use. All the ingredients other than carrot and cashewnut were mixed thoroughly in a mixer. Dried carrot strips and cashewnuts were then added and mixed gently.

C. Halwa Mix

Halwas play an important role among sweets. Halwas are of middle Eastern origin and the art of making it has spread all the way from the Mediterranean sea to the Bay of Bengal. Halwas are prepared from cereals, vegetables or fruits with sugar, fat, with or without milk solids, flavouring agents and nuts. An enriching whole wheat halwa filled with nuts and raising is very popular in South India. Halwas are generally rich in nutrients and calories. Many of the them make good supplements of our diet. When vegetables were used for the preparation of halwas, they provide a concentrated, source of vitamins and minerals. Carrot and gourd halwas are frequently made for banquets and feasts.

In many of the countries, halwa is consumed as a dessert served usually along with the meal or after the meal. Halwa is liked by all age groups and it constitutes a main source of calories and nutrients. Halwas belongs to intermediate moisture foods as a moisture content ranged between 10 and 50%. The shelf life of foods can be increased by decreasing the moisture content. The BIS (1975) specification for Bombay Halwa is 12.0% moisture, 3.0% acidity, 55% sucrose and 6.0% fat.

Preparation of Halwa Mixes

A. Carrot Halwa Mix

Ingredients: Dehydrated carrot-100 gm, milk powder – 50 gm, sugar – 150 gm, cardamom powder – 2 gm.

Method: The steps involved were as follows:

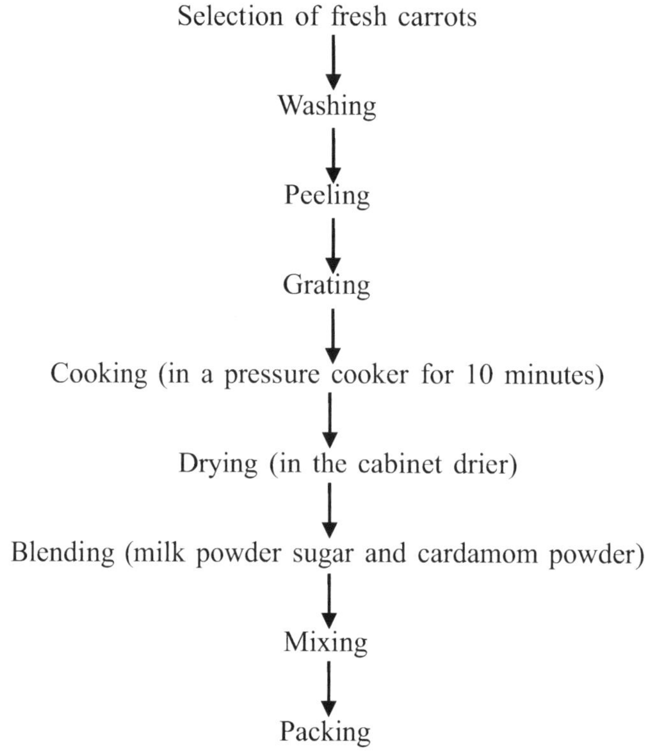

Figure 14: Flow chart for the preparation of carrot halwa mix

B. Bombay Halwa Mix

Ingredients: Maida flour – 80 gms, corn flour –20 gms, Kesari colour – little.

Method: The ingredients were dried in a cabinet drier at 80°C for 2 hrs. Then it was cooled and all the ingredients were mixed thoroughly and packed in a polythene bag.

C. Gulabjamun Mix

Gulabjamun is a popular Indian sweet meat traditionally prepared from khoa. Gulabjamun mix is prepared by blending milk powder / khoa with maida and baking powder.

Preparation of Gulabjuman Mix	:	**Ingredients**
Milk powder / khoa	:	200 gm
Maida	:	20 gm
Baking powder	:	1 gm

Method: The ingredients were dried in a cabinet drier at 80°C for 2 hrs. Then it was cooled and all the ingredients were mixed thoroughly and packed in a polythene bag.

II. Preparation of Jangiri Mix, Cake Mix and Ice Cream Mix

A. Jangiri Mix

It is a traditional confection made from wheat flour, deep fried and soaked in sugar syrup and eaten as snack particularly on festive occasions. Originated in Indian subcontinent production carried by Indian immigrants to Europe, N. S. America and West India and South East Asia.

Ingredients need

Maida flour	:	50 gm
Curd flour /lentil flour	:	50 gm
Sodium bi carbonate	:	2 gm
Citric acid	:	2 gm
Yellow colour	:	0.5 gm

Method

The ingredients were dried in the cabinet drier at 80°C for 2 hrs. Then it was cooled and all the ingredients were mixed thoroughly and packed in a polythene bag.

B. Cake Mix

Ingredients needed

Maida flour	:	100 gm
Skim milk powder	:	50 gm
Powdered sugar	:	75 gm
Baking powder	:	2.0 gm
Sodium bi carbonate	:	2.0 gm
Gluten	:	2.0 gm

Method: The weighed quantum of maida (100 gm) was mixed in a stainless steel vessel using a ladle followed by the addition of sugar. Finally, the baking powder, sodium bicarbonate, gluten and skim milk powder were added slowly to the content and mixed thoroughly. For uniform mixing, the processed instant mix was passed through the sieve (BS 60).

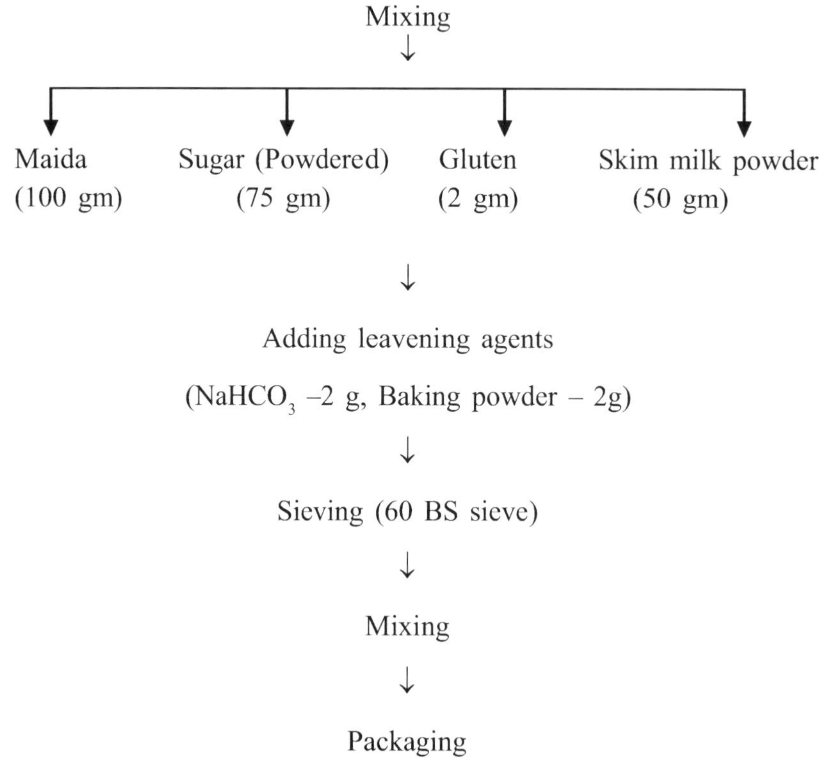

Figure 15: Flow chart for cake ready mix

C. Ice Cream Mix

Ice cream is most palatable and nourishing food which was once considered to be a sophisticated item, which is becoming more and more popular among all sections of the people. The popularity and higher consumption of the ice cream could be attributed to its refreshing cool and delightfully sweet characteristics. Ice cream is composed of a mixture of food materials such as milk products, sweetening materials, stabilizers, flavour or egg products which are referred to as ingredients. The composition of ice cream is usually expressed as percentage of its constituents. It contains 10%, fat, 11%, SNF, 15% sugar and 0.3 % stabilizer. The physical characteristics such as viscosity, whipping ability, freezing parameters, texture, melt down rate, crystal formation and sensory characteristics were used for the quality evaluation of ice cream apart from chemical and microbiological analysis.

Preparation of Instant Ice Cream Mix	:	**Ingredients**
Corn flour	:	100 gm

Sugar	:	788 gm
Skim milk powder	:	100 gm
Stabilizer (gelatin)	:	6 gm
Emulsifier (Glycerol mono sterate)	:	6 gm
Total	**:**	**1000 gm**

Method: The corn starch was mixed with stabilizer and emulsifier thoroughly. The powdered sugar was added little by little to the content (starch mix) by mixing continuously. Finally the skim milk powder was added to it and the instant ice cream mix was folded gently with the help of ladle. For uniform mixing the processed ice cream mix was passed through the sieve (BS 60). The instant ice cream mix was packed in MPP pouches and stored at room temperature.

Quality Evaluation of Traditional Sweet Mixes

Preparation of Sweets

(a) Rice kheer: About 750 ml of water was heated to tapid condition (40-45°C). The instant kheer mix (100 gm) was added to it. The mixture was stirred continuously to prevent lump formation and cooked. The kheer will be ready for serving within seven minutes.

(b) Carrot kheer: The carrot kheer mix (100 gm) was soaked in (reconstitution) boiling water (200 ml) for 3 min. and served hot.

(c) Carrot halwa: To 50 ml of hot water, 50 gm of instant carrot halwa mix was added, mixed and heated for 5 minutes, to obtain halwa. The cooked mass was poured into a greased plate, allowed to cool and cut into desired shape.

(d) Bombay halwa: For one measure of the mix (100 gm), add two measures of water (200 ml), 2 measures of sugar (200 gm) and mixed well. Cook the mix after adding dalda (or) ghee (10 ml) for 20 minutes by stirring continuously. At the end add the broken cashewnuts (5 gm), and melon seeds (5 gm). Pour into a greased plate, allow it to cool and cut into pieces.

(e) Gulabjamun : Ingredients: Gulabjamun mix – 100 gm, water – 15 ml.

For the preparation of sugar syrup: Water – 300 ml, sugar – 300 gm,
For frying: Oil – 100 ml.

Method

1. Smooth dough was prepared by mixing water (15 ml) and gulabjamun mix.
2. Dough was made into small round balls (approximate weight – 5 gm).
3. Oil was heated to 180°c and the balls were deep fat fried at low flame for 5-7 min.

Preparation of Sugar Syrup

1. The sugar was taken in a vessel with water.
2. The content was heated by stirring till the sugar syrup reaches to 50° brix.
3. The prepared syrup was filtered.
4. The fried gulabjamun was transferred to the hot syrup and allowed to soak for 60 minutes.
5. Sugar syrup impregnated gulabjamun was served with syrup (each gulabjamun with two table spoons of syrup).

(f) Cake: The other ingredients required at the time of preparation of cake from instant cake mix are milk – 50 ml and butter – 50 gm.

Method

1. Butter and milk (30 ml) were blended (creamed) continuously.
2. Cake mix was added little by little to the cream with continuous stirring.
3. More (20 ml) milk was added to have batter consistency.
4. The baking cups were greased (50 gm cup).
5. The batter was poured into the greased cup to $1/4^{th}$ of the cup height.
6. Cakes were baked in the prepared oven at 180°C for 15 minutes.

(g) Ice cream: The other ingredients required at the time of preparation of ice cream from instant ice cream mix are (1) milk – 500 ml, fresh cream – 60 gm and liquid glucose – 15 gm.

Method

1. Five hundred ml of milk was taken in a heavy bottom vessel and heated to boil.
2. The instant ice cream mix was made into paste using raw milk and added to the boiling milk followed by liquid glucose and stirred continuously for 2 minutes.
3. The content was cooled and fresh cream was added and mixed again.
4. The content was whipped in a mixie for 30 min. before freezing.
5. The whipped sample was poured in a ice cream tray or cup and kept aside the freezer for 4 hours (freezing temperature – 20°C).
6. To obtain smooth and soft ice cream, the content was whipped at every ½ hour during the course of freezing.
7. The prepared ice cream was stored at – 20°C in the freezer.

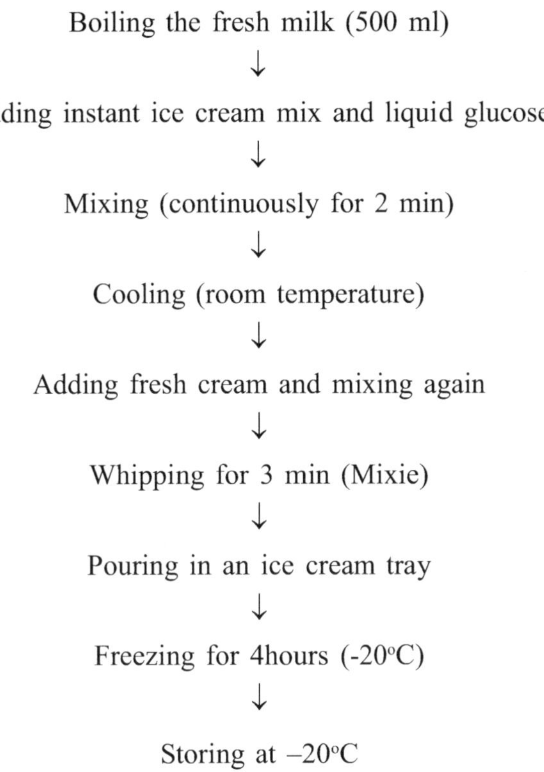

Figure 16: Flow chart for the preparation of ice cream

References

Austin, A. and A.Ram, Studies on Chapati- Making Quality of Wheat, Indian Council of Agricultural Research, New Delhi, 1971.

Bennion, Marian: Introductory Foods, 8th edn., Macmillan Publishing Co.Inc., New York, 1985.

Boersen, A.C. 1990. Spray drying technology. J. Soc. Dairy Technol. 43(1), 5-7

Charm, S.E. 1978. Dehydration of foods. In The Fundamentals of Food Engineering, S.E. Charm (Editor). AVI Publishing Co., Westport, CT, pp. 298-408

Erickson, L.E. 1982. Recent developments in intermediate moisture foods.J.Food Protech. 45(5), 484-491

Fact, R.B. and Caldwell, E.F. 1990. Breakfast Cereals and How they are made. American Association of Cereal Chemists, St. Paul, MN.

Frazier W.C. and D.C. Westhoff: Food Microbiology, 3rd edn, McGraw-Hills, New York, 1988.

Peckham, Gladys, C. and Jeanne H.: Freeland –Graves. Foundations of Food Preparation, 5th Edn., Macmillan, New York, 1987.

Puruthi, J.S., Spices and condiments, National Book Trust of India, New Delhi, 1976.

Taoukis, P.S., Breene, W.M., and Labuza, T.P.1987. Intermediate moisture foods. Adv. Cereal Sci. Technol. 9, 91-128.

Van Arsdel, W.B., Coppley, M.J., and Morgan, Jr., A.I. 1973. Food Dehydration. 2nd ed. Vols. 1 and 2. AVI Publishing Co., Westport, CT.

❑❑❑

Chapter - 3

Fruits and Vegetables Based Foods

Importance of Fruits and Vegetables in Human Life

Fruit is the edible, and more or less juicy, product of a tree or plant and consist of the matured ovary including its seeds and adjacent parts. Usually fruits are sweet, with a wide range of flavours, colours and textures.

Composition

Fruits are very poor source of protein and fat. Avocado is the exception containing 28% fat. Fruits contain high amount of moisture hence they are highly perishable. They are also good source of fiber. Fruits are not very good sources of calories. Fruits like bananas give fairly good amount of calories. Ripe fruit contains a higher percentage of sugar than unripe fruit does and the sugar is chief in the form of sucrose, fructose and glucose. Generally fruits are poor source of iron, Mangoes are excellent source of carotenes. Oranges are fairly good source of beta carotene. Guavas are the best source of vitamin C. Citrus fruits are good source of vitamin C. Cashew fruits are inexpensive and rich in vitamin C, although there is variation of vitamin content from fruit to fruit most fruits in the raw state contain some ascorbic acid. If fruits are bruised, peeled, cooked or exposed to air, large amounts of the vitamin may be oxidized. Apples are not only expensive; they contribute little to the nutritive value. They give fiber to the diet.

Value Added Products from Fruits

Juice, RTS, nectar, squash, cordial, jam, preserve, toffee, amchur, pickle, chutney, canned product, fruit powder, concentrate, jelly, cheese, toffee, vinegar, syrup, candy, wine, dried product, marmalade, cider, pickle.

- Fruit Beverage
- Jam, Jelly, Marmalade
- Candy

- Preserve
- Dehydration of Fruits and Pickle Production

A wide range of plant materials are used to manufacture beverages. These include leaves, stems, sap, fruits, tubers, and seeds (grains).

The following list may prove helpful in distinguishing between the different types of drink:

- ***Juices.*** These are pure fruit juice with nothing added.
- ***Nectars***. These normally contain 30 per cent fruit solids and are drunk immediately after opening.
- ***Squashes.*** These normally contain at least 25 per cent fruit pulp mixed with sugar syrup. They are diluted, to taste, with water and may contain preservatives.
- ***Cordials***. These are crystal-clear squashes.
- ***Syrups***. These are concentrated juices which are clear. They normally have a high sugar content.

Each of the above products is preserved by its natural acidity and by pasteurization. Some drinks (syrups and squashes) also contain a high concentration of sugar which helps to preserve them.

Mango *(Mangifera Indica)*

In general harvest maturity in mango is reached in 12-16 weeks after fruit set, depending on variety. Specific gravity is a good criterion to judge maturity of the fruit. Mangoes are harvested by hand if the pickers can reach them. Fruits on high branches are harvested with a picking pole having a cloth bag and cutting knife at the top. Fruits are to be harvested with little stalk to prevent latex trickling which leads to stem end rot. Mangoes are generally harvested at physiological mature stage and it takes 6-14 days to ripen under ambient conditions.

The ripening phenomenon is associated with conversion of starch to sugars and loss of firmness of fruit. Mango is one such fruit, which can be processed at almost every stage of growth, development, maturity and ripening. Raw mango fruits are utilized for mango powder, pickle, chutney etc. An excellent drink can also be made from green mangoes. Ripe mangoes are utilized for making slab, toffee various beverages such as nectar, squash etc. Drying after exposing to sulphur fumes also preserves ripe mango slices. Methods have also been standardized to produce cryogenically (liquid N) frozen mango slices.

Guava *(Psidium Guajava)*

Guava fruits re most commonly harvested by hand. Firm yellow to half-yellow mature fruits are harvested. Over ripe fruits are easily damaged in transport and handling. Fruits that are immature when harvested do not develop into quality ripe fruits. Guava fruits can be kept in ventilated polyethylene bags for 10 days at ambient temperature 18-20° C Guava is very popular as a fresh fruit because of its excellent taste, high vitamin content and 100% edibility. This fruit is equally important for the processing industry.

A large number of processed products are manufactured from guava. Because of presence of rich amount of pectin, a high quality natural jelly is obtained from guava. Processed guava pulp is an excellent raw material for preparation of various other guava products such as nectars, beverages, jams, toffee, cheese, ice cream topping etc. Guava pulp can be preserved successfully in bulk either by application of heat aseptic packaging or addition of chemical preservation (SO2). Cannel guavas with sugar syrup (40° Brix), dehydrated guavas, and guava powder are the other important products.

Pomegranate *(Punica Granatum)*

Since, pomegranate is highly popular as a fresh fruit; it is not used for processing to a great extent. Pomegranate juice is highly acceptable drink. The steps in making of juice drink include, extraction of juice, clarification in a flash pasteurizer, cooling, settling for 24 hours, racking up, filtering heat preservation. Anardana is made from pomegranate seeds, particularly of the sour type, after drying that we are used as acidulant for culinary purpose.

Custard Apple *(Annona Squamosa)*

Custard apple is harvested in several installments, but the best harvesting stage is when the firm fruit begins to develop colour. It is generally picked when it becomes creamy yellow between the segments and begins to crack slightly. The fruit has the tendency to burst open if kept on the tree for a long time. Custard apple is highly perishable and cannot be stored for long time. It can be stored successfully for 9 weeks at 7-10° C with 85% to 95% RH. Lower storage temperature induces chilling injury. Custard apple can be kept for 9 days after treating with 50 ppm Bavistin and placing in a polythene bag containing Kmno4 compared to untreated fruits for 5 days. Custard apple is not used for processing purpose to a great extent.

On heating the pulp at develops bitterness. To pulp can be frozen successfully for use in the ice cream industry. Ready to serve beverages are made from custard apple. Bitterness of the pulp can be removed by treating with peptic enzyme. The pulp can be frozen successfully for use in ice-cream industry. At CRIDA, a simple technique has been developed for manual extraction of custard apple pulp by rotatory motion of a round hair comb in the scooped fruit held in stainless steel sieve. This pulp can be supplied to the ice-cream industries.

Jackfruit *(Artocarpus Heterophyllus)*

The fruit is used both in the unripe and ripe stage. Raw jackfruit is popularly used as a vegetable. Fully mature but unripe fruits are harvested and appearance and a dull sound upon tapping judge fruit maturity. Ripe jackfruit is consumed as a dessert fruit. Jackfruit chips are prepared by frying ripe or semi-ripe fruits. Jackfruit leather is also prepared from the ripe or semi-ripe fruit. A palatable beverage concentrate can be made from jackfruit pulp by adding sugar, citric acid and water. In addition high class canned, frozen and dried products such as nectar, preserves confections etc can be prepared from the ripe fruits. The green jackfruit utilized for making pickle, canned and curried vegetables. The wastes (skins, peels and cores), which constitute about 45% of the total fruit weight, have been found to be a fairly good source of pectin.

Bael fruit *(Aegle Marmelos)*

The bael fruit is known for its medicinal properties. The bael fruit is one of the most nutritious fruits. It contains 61.5 g of water, 1.8 g of protein, 1.7 g of minerals, 31.8 g of carbohydrates and 1.19 mg of riboflavin/100 g edible portion. It may be noted that no other fruits has such a high content of riboflavin. Bael fruit has been used from time immemorial for processing in the mature green form to prepare preserves. The difficulty in the extraction of ripe bael fruit pulp is overcome by addition of water equal in weight to the pulp, adjusting the pH to 4.3 with citric acid and heating at 80° C for one minute, before passing through the extractor/pulper.

Addition of water dilutes the mucilage and the Manual extraction of Custard apple pulp application of heat rot only inactivates the enzymes but also helps in dissolving the mucilage uniformly throughout the pulp. The fruit pulp thus obtained has almost the same consistency and colour as mango pulp. Ripe bael fruit pulp, if extracted properly can be used for the preparation of various fruit products viz., nectar squash/leather/ slab, powder etc., which can be commercially exploited.

Aonla *(Phyllanthus Emblica)*

The fruit is highly nutritious and is a rich source of pectin and polyphenols apart from ascorbic acid. Aonla fruits are well known for their medicinal properties. The fruits are used for curing chronic dysentery, bronchitis, and diabetes. The storage of Aonla depends on maturity at harvest. The fruit keeps well in cool chamber for 17-18 days compared to 8-9 days at ambient temperature. Aonla fruit is seldom consumed fresh but the fruit is valued highly in the Ayurvedic system of medicine.

In Ayurvedic preparation like 'Chyavanprash' and triphala, Aonla is one of the main ingredients. Fruit products like pickle, preserve, candy, jam, syrup and dried shreds are made from Aonla. Aonla preserve is very important article of commerce and is in great demand. Streaming or blanching the fruit prior to processing can minimize ascorbic acid loss in the products. It is also used in tanning and dyeing industries.

A technique has been developed at CRIDA for separation of segments of aonla and does away with nut by steaming. These segments were used for preparation of different products. In a study on the suitability of different varieties for processing into candy, murabba and pickle, the candy of variety Chakkiya ranked first in respect of ratings for color, flavor, texture and total score and overall ranking. The variety Banarasi ranked first for texture and variety, Francis ranked first for color. After a storage period of 4 months at room temperature, the variety Chakkiya again ranked first in all attribute except flavor. The candy of Banarasi ranked second overall. In case of murabba the variety Krishna ranked first in all attribute followed by Chakkiya, which ranked second in all attribute except texture, where it ranked first.

Squash was prepared by blending Aonla juice with other juices viz., ginger, roselle, pineapple and lime. Organoleptic evaluation of squash revealed that score for color, flavour and consistency increased with addition of ginger and Roselle. The blend of aonla, ginger, roselle (80:15:5) ranked first in all attributes including overall ranking, this was followed by blend of aonla, lime, ginger (75:20:5). Roselle helps in improving the color and ginger helps in improving the flavor of squash.

Ber (*Zizyphus Mauritiana*)

Ber fruit are consumed as such or can be processed into different fruit products. Juicy varieties are better suited for pulp and juice extraction. The fully ripe, well-developed fruits are washed de-stoned and juicer extracts juice. Ber juice can be used for the preparation of ready to serve beverage. Carbonated beverage of ber is highly acceptable and has excellent keeping quality. Dehydrated for is prepared by treating ber fruits with sulphur dioxide at 3.5-10 g/kg for 3 hours followed by sun drying, or carbinet drying below 15% moisture. Ber can be utilized for candy and ber pulp can be processed into wine. The steps include diluting the pulp, adding pectinase enzymes adjusting proper Brix with sugar, addition of yeast, fermentation, stabilization and clarification.

Jamun (*Syzygium Cumini*)

It is reported that jamun fruits are used for making products such as jam, jelly beverages, wine and vinegar. It has been found that maximum yield of jamun juice with a high level of anthocyanins and other soluble constituents can be obtained by grating the fruit, heating to 70°C, and passing through basket press. The jamun juice thus obtained is again heated to 85°C and then cooled to room temperature. Sodium benzoate (500 pm) is added to the juice before it is stored. Pure jamun juice can also be stored by heat pasteurization. The juice being highly acidic is not consumed as such. A ready to serve beverage (nectar) is prepared with 25% juice, 18° Brix and 0.6% acidity. Jamun seeds are also known for their properties which help to cure diabetes, diarrhea and dysentery.

Phalsa *(Grewia Subinaequalis)*

Being highly perishable, the fruit must be utilized within 24 hours after picking. The popularity of phalsa fruit is due to its attractive colour ranging from crimson red to dark purple and its pleasing taste. The juice when extracted gives a deep crimson red to dark purple colour and is very popular. It is rated very high in indigenous system of medicine. The juice is extremely refreshing and is considered to have a cooling effect especially in hot summer. Heating the crushed phalsa fruit to 50°C gives the highest recovery of the juice with an appropriate quantity of anthayanin and other soluble and insoluble materials. Studies have shown that addition of cane sugar to the juice has a protective effect on colour stability. Aonlla squash blended with other juices

Roselle

The calyces of Roselle (*Hibiscus sabdarifa L*) which grows well even when sown in September under rainfed conditions in Alfisols can be utilized for processing into different products. It can be a very good source for coloring fruit products that don't have attractive color. With the ban on coal tar food dyes, it is easy to arouse interest in Roselle as a coloring source. Today, Roselle is attracting the attention of food and beverage manufactures and pharmaceutical concerns who feel it may have exploitable possibilities as a natural food product and as a colorant to replace some synthetic dyes.

- Chopped and added to fruit salads.
- Stewing as sauce.
- Roselle syrup.
- No need to add pectin to jelly. In fact the calyces posses 3.19% pectin and in Pakistan Roselle has been recommended as a source of pectin for the fruit preservation industry.
- Juice made by cooking a quantity of calyces with ¼ water in ratio to amount of calyces is used for cold drinks and may be frozen or bottled if not for immediate needs. In West Indies and America (tropical) Roselle is prized primarily for the cooling lemonade like beverage made from calyces.
- After extraction of color, acids and pectin for jelly making, the calyces mass could be further processed to a sauce which was organoleptically highly acceptable.

Guava

Jellies with Guava – Roselle Blends

Calyces of roselle can be used as a source of colorant for guava jelly, since there is a problem of browning in guava jelly during storage. Organloeptic evaluation of jellies prepared from different proportions of guava – roselle (calyces) revealed the preference of tasters for blending at 85:15 proportions to get jelly with attractive color. Roselle calyces can be used with guava for jelly making without significant change in guava flavor.

Value Addition and Processing of Fruit and Vegetables

Amla processing and value addition are engaging different units as follows

- Amla Washing Unit
- Amla Shredder (for removal of amla seed)
- Pulverizer or Pulper
- Hydraulic Basket Press (Juice Section)
- Centrifuge type Filtration Unit (juice Section)
- Steam Jacketed Kettle (Juice section)
- Homogenizer (Cosmetics and oil section)
- Syrup Preparation Pan (Candy Section)
- Mixing Pan (Candy Section)
- Oil Preparation Vessel (Oil Section)
- Sparkler Filter (Oil Section)
- Tray Dryer (Powder, Supari and candy section)
- Pulverizer (Powder section)
- Wood Fired Boiler

Figure 17: Amla processing machineries

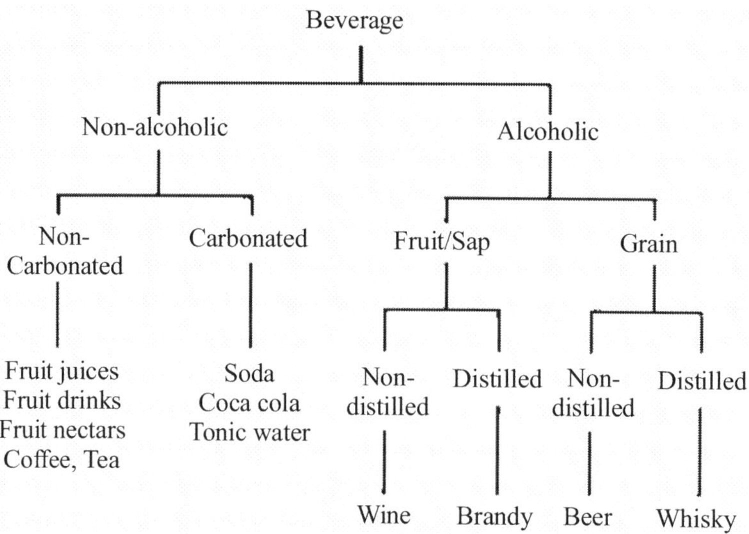

Figure 18: Flow diagram of Beverages classification

The market for beverages is broadly divided in many countries into those products that are bought to quench thirst, and those that are consumed on special occasions including festivals. The former group are mostly nonalcoholic and include tea, coffee, and soft drinks (including juices, nectars, and carbonated drinks). In some countries these products are also used on social occasions, whereas in other areas alcoholic beverages are preferred (although soft drinks are also usually available). In most countries, the market for alcoholic and non-alcoholic drinks is specific with regard to religious and cultural taboos.

Competition from medium/large-scale producers is most acute for small-scale producers in beverage manufacture. Many large-scale producers promote their products by implying status in their consumption and spend considerable amounts on advertising and packaging. They may also have established sophisticated distribution systems and specific agreements with wholesalers and retailers. Thus beverage manufacture is one of the most difficult for small-scale producers to establish and succeed in.

Nutritional Significance

Most beverages contain a great deal of water. This does not add many nutrients to the diet, but it does play an important role in maintaining body balance by preventing dehydration.

Beverages are not usually consumed for their food value, but many, particularly the fruit drinks, contain quite a high percentage of sugar and therefore add to the energy content of the diet. Additionally fruit juices provide a supply of vitamins and minerals.

Certain drinks contain artificial flavourings and colourings. The use of such additives is governed by legal requirements and it is vital to keep to these regulations in order to protect the consumer from any undesirable side-effects. Some colouring agents for example are thought to cause hyperactivity in children, and are therefore to be avoided.

Alcoholic drinks are judged in terms of flavour and the stimulant effect they produce. In many countries alcohol production is strictly controlled by government agencies and it may be difficult to obtain the necessary permits to produce these beverages legally.

Non-Alcoholic Beverages

A wide range of drinks can be manufactured which contain as the base material either pulped fruit or juice. Many are drunk as a pure fruit juice without the addition of other ingredients, whereas others are diluted with sugar syrup.

For simplicity, fruit drinks can be divided into two groups:

- Those that are drunk immediately after opening.
- Those that are used little by little from bottles which are stored between use.

The former group should not need any preservative if processed and packaged properly. However the latter must contain a certain amount of permitted preservatives to have a long shelf-life after opening.

Fruit Beverages

Fruit beverages are easily digestible, highly refreshing, thirst quenching, appetizing and nutritionally far superior to many synthetic and aerated drinks. They can be classified into unfermented and fermented beverages.

Unfermented Beverages

Fruit juices which do not undergo alcoholic fermentation are termed as unfermented beverages. They include natural and sweetened juices, RTS, nectar, cordial, squash, crush, syrup, fruit juice concentrate and fruit juice powder. Barley waters and carbonated beverages are also included in this group.

Tropical fruits are now considered as an important item of commerce as they have gained enormous market potential. Post-harvest losses of fruits and vegetables are more serious in developing countries than those in well-developed countries. The total losses from harvest to the consumer point are as high as 30-40%, which is worth thousands of crores of rupees. About 10-15% of fresh fruits and vegetables shrivel and stale, lowering their; market value and consumer acceptability. Minimizing these losses can increase their supply. It will also keep pollution under control. Improper handling and storage cause physical damages due to tissue breakdown. Mechanical losses include bruising, cracking cuts, microbial spoilage by fungi and bacteria, whereas physiological losses include transpiration, pigments, organic acids and flavour. About 30% of fruit decay due to Penicillium spices.

Tropical fruits, which are at present under-utilized, have an important role to play in satisfying the demand for nutritious, delicately flavoured and attractive natural foods of high therapeutic value. They are in general are accepted as being rich in vitamins, minerals and dietary fibre and therefore are an essential ingredient of a healthy diet (Table). Apart from nutritive, therapeutic and medicinal values quite a few of these tropical fruits have excellent flavour and very attractive colour. Fruits like jamun and phalsa are highly perishable. Bael fruit is not an easy to eat out of hand item. Many people because of its strong astringent taste do not like fresh Aonla fruit. However, these fruits have unlimited potential in the processed form and consumers all over world can get opportunity to enjoy the fruits in the form of their processed products.

Processing

Processing is the best way of utilizing surplus production of fruits during seasonal gluts.

Advantages of Processing

- Helps in converting perishable fruits in to durable form
- Fruits, which are very difficult to eat out of hand, can be processed in to a range of highly acceptable fruit product.
- Helps in reducing wastage.
- Value addition.

Methods of processing of fruits into products

- Preservation by heat treatment.
- Asceptic packaging.
- Preservation of by removal of heat.
- Quick freezing.
- Preservation by removal of moisture.
- Preservation by addition of chemicals.
- Minimal processing.

I. Fruit beverage

A. Squash

This is a type of fruit beverage containing at least 25 per cent fruit juice or pulp and 40 to 50 per cent total soluble solids, commercially. It also contains about 1.0per cent acid and 350 ppm sulphur dioxide or 600 ppm sodium benzoate. It is diluted before serving.

Mango, orange and pineapple are used for making squash commercially. It can also be prepared from lemon, bael, papaya, etc. using potassium metabisulphite (KMS) as preservative or from jamun, passion-fruit, peach, plum, raspberry, strawberry, grapefruit, etc. with sodium benzoate as preservative.

Fruits

Washing

Trimming

Cutting or grating

Juice extraction

Straining

Juice measuring

Preparation of syrup
Sugar + water + acid, heating just
to dissolve)

Straining

Mixing with juice

Addition of preservation
(0.6 g KMS or 1.0 g sodium
Benzoate / litre squash)

Bottling

Capping

Storage

Figure 19: Flow diagram for preparation of squash

B. Ready-to-Serve (RTS)

This is a type of fruit beverage which contains at least 10 per cent fruit juice and 10 per cent total soluble solids besides about 0.3 per cent acid. It is not diluted before serving, hence it is known as ready-to-serve (RTS).

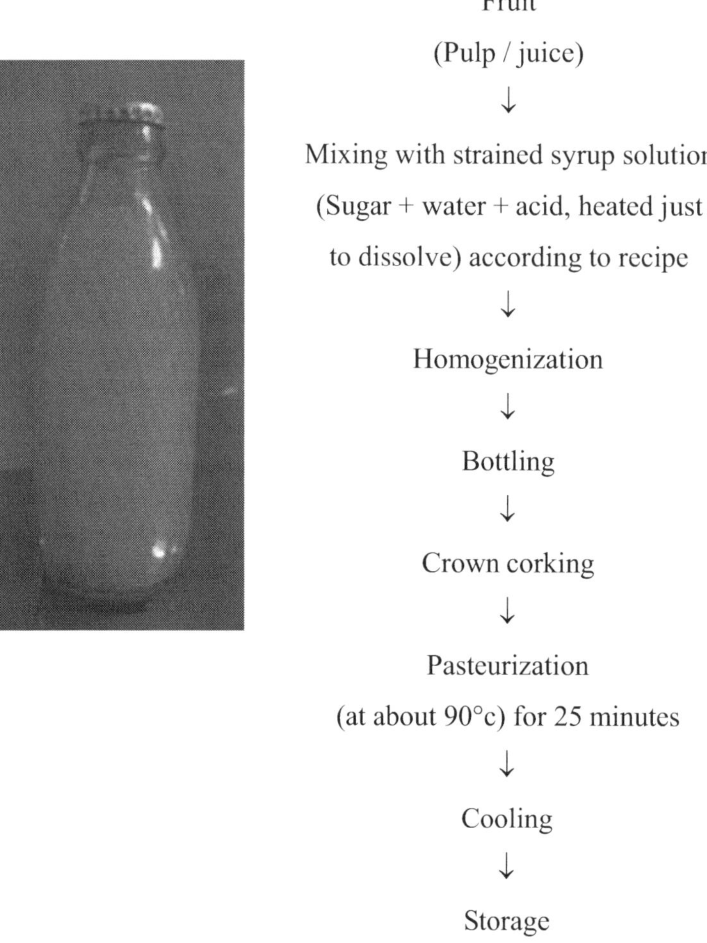

Fruit

(Pulp / juice)

↓

Mixing with strained syrup solution

(Sugar + water + acid, heated just

to dissolve) according to recipe

↓

Homogenization

↓

Bottling

↓

Crown corking

↓

Pasteurization

(at about 90°c) for 25 minutes

↓

Cooling

↓

Storage

Figure 20: Flow diagram for processing of RTS beverages

C. Cordial

It is a sparkling, clear, sweetened fruit juice from which pulp and other insoluble substances have been completely removed. It contains at least 25 per cent juice and 30 per cent TSS. It also contains about 1.5 per cent acid and 350 ppm of sulphur dioxide. This is very suitable for blending with wines. Lime and lemon are suitable for making cordial.

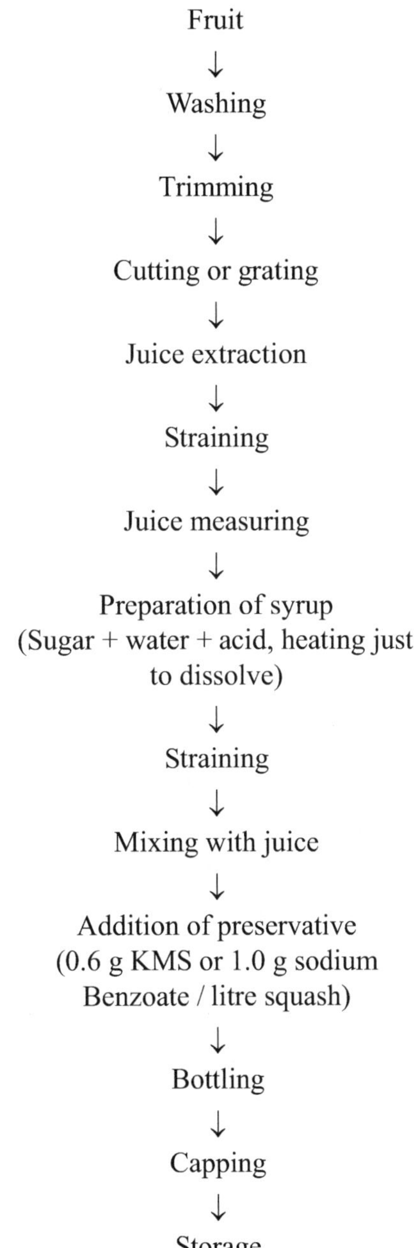

Figure 21: Flow diagram for processing of cordial

D. Nectar

This type of fruit beverage contains at least 20 per cent fruit juice / pulp and 15 per cent total soluble solids and also about 0.3 per cent acid. It is not diluted before serving.

Table 20: Preparation nectar from different fruits

S. No.	Fruit	Juice / Pulp (%)	Quantity of water required (liter)
1.	Mango	20	Quantity of finished product (liter) – Quantity of (juice (liter) + sugar (kg) + acid (kg) used
2.	Papaya	20	
3.	Guava	20	
4.	Bael	20	
5.	Jamun	20	
6.	Aonla (blend)	Aonla pulp 20 Lime juice 2 Ginger juice 1	

For preparing the above beverages, the total soluble solids and total acid present in the pulp/juice are first determined and then the requisite amounts of sugar and citric acid dissolved in water are added for adjustment of TSS and acidity.

II. Fermented Beverages

Fruit juices which have undergone alcoholic fermentation by yeasts include wine, champaigne, port, sherry, tokay, muscat, perry, orange wine, berry wine, nira and cider.

III. Jam, Jelly and Marmalade

A. Jam

Jam is a product made by boiling fruit pulp with sufficient amount of sugar to a reasonably thick consistency, firm enough to hold the fruit tissues in position. Apple, pear, sapota (chiku), peach, papaya, karonda, carrot, plum, straw-berry, raspberry, mango, tomato, grapes and muskmelon are used for preparation of jams. It can be prepared from one kind of fruit or from two or more kinds

B. Jelly

A jelly is a semi-solid product prepared by boiling a clear, strained solution of pectin-containing fruit extract, free from pulp, after the addition of sugar and acid. A perfect jelly should be transparent, well-set, but not too stiff, and should have the original flavour of the fruit. It should be of attractive colour and keep its shape when removed from the mould. It should be firm enough to retain a sharp edge but tender enough to quiver when pressed.

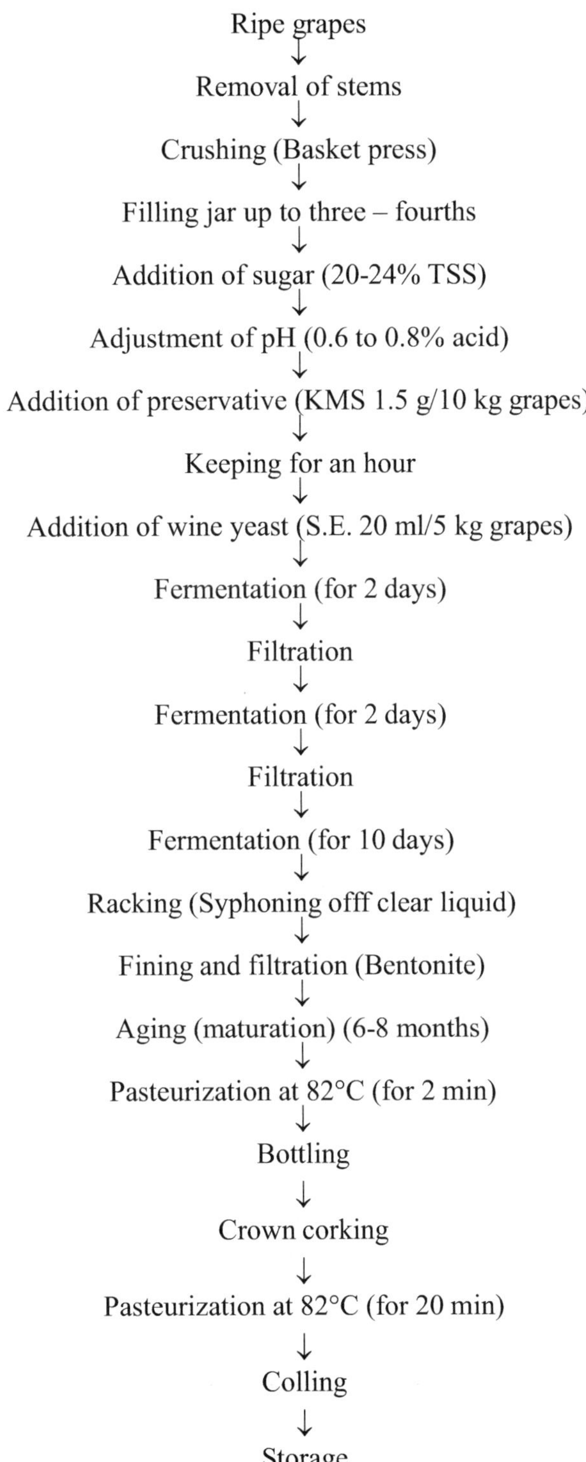

Figure 22: Flow diagram for processing of Wine

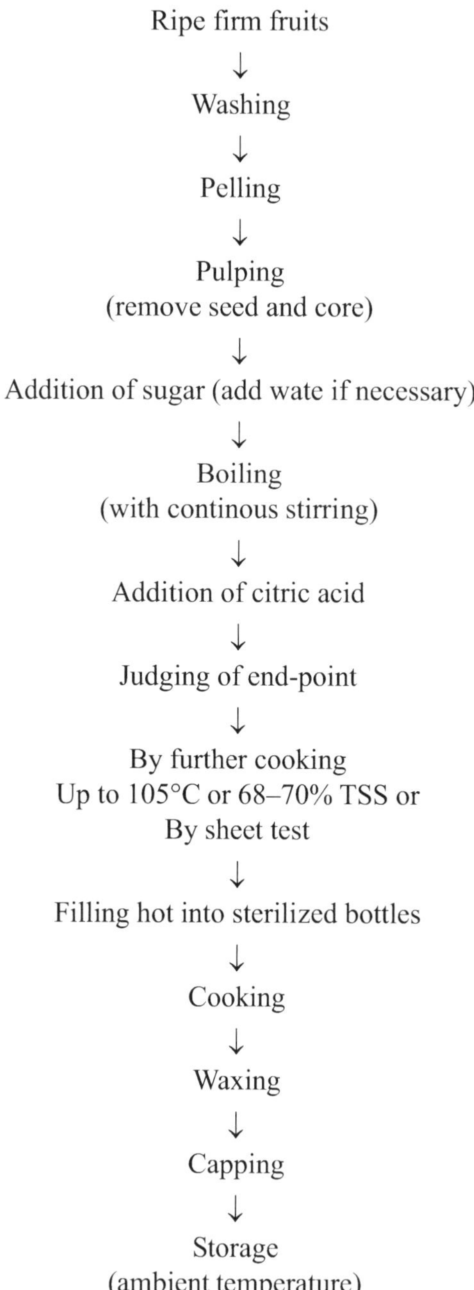

Figure 23: Flow diagram for processing Jam

Guava, sour apple, plum, karonda, wood apple, loquat, papaya and goose-berry are generally used for preparation of jelly. Apricot, pineapple, strawberry, raspberry, etc. can be used but only after addition of pectin powder, because these fruits have low pectin content.

C. Marmalade

This is a fruit jelly in which slices of the fruit or its peel are suspended. The term is generally used for products made from citrus fruits like oranges and lemons in which shredded peel is used as the suspended material. Citrus marmalades are classified into (i) jelly marmalade, and (ii) jam marmalade.

I. Jelly Marmalade

The following combinations give good quality of jelly marmalade:

1. Sweet orange (Malta) and khatta or sour orange (Citrus aurantium) in the ratio of 2:1 by weight. Shreds of Malta orange peel are used.
2. Mandarin orange and khatta in the ratio of 2:1 by weight. Shreds of Malta orange peel are used.
3. Sweet orange (Malta) and galgal (***Citrus limonia***) in the ratio of 2:1 by weight. Shreds of Malta orange peel are used.

II. Jam Marmalade

The method of preparation is practically the same as that for jelly marmalade. In this case the pectin extract of fruit is not clarified and the whole pulp is used. Sugar is added according to the weight of fruit, generally in the proportion of 1:1. The pulp-sugar mixture is cooked till the TSS content reaches 65 per cent.

III. Candy

A whole fruit / vegetable or its pieces impregnated with cane sugar or glucose syrup, and subsequently drained free of syrup and dried, is known as candied fruit / vegetable. The most suitable fruits for candying are aonla, karonda, pineapple, cherry, papaya, apple, peach, and peels of orange, lemon, grapefruit and citron, ginger, etc.

The process for making candied fruit is practically similar to that for preserves. The only difference is that the fruit is impregnated with syrup having a higher percentage of sugar or glucose. A certain amount (25-30 per cent) of invert sugar or glucose, viz., confectioners' glucose (corn syrup, crystal syrup or commercial glucose), dextrose or invert sugar is substituted for cane sugar. The total sugar content of the impregnated fruit is kept at about 75 per cent to prevent fermentation. The syrup left over from the candying process can be used for candying another batch of the same kind of fruit after suitable dilution for sweetening chutneys, sauces and pickles and in vinegar making.

Glazed Candy

Covering of candied fruits / vegetables with a thin transparent coating of sugar, which imparts them a glossy appearance, is known as glazing. Cane sugar and water (2:1 by weight) are boiled in a steam pan at 113-114°C and the scum is removed as it comes up. Thereafter the syrup is cooled to 93°C and rubbed with a wooden ladle on the side of the pan when granulated sugar is obtained. Dried candied fruits are passed through this granulated portion of the sugar solution, one by one, by means of a fork, and then placed on trays in a warm dry room. They may also be dried in a drier at 49°C for 2-3 hours. When they become crisp, they are packed in airtight containers for storage.

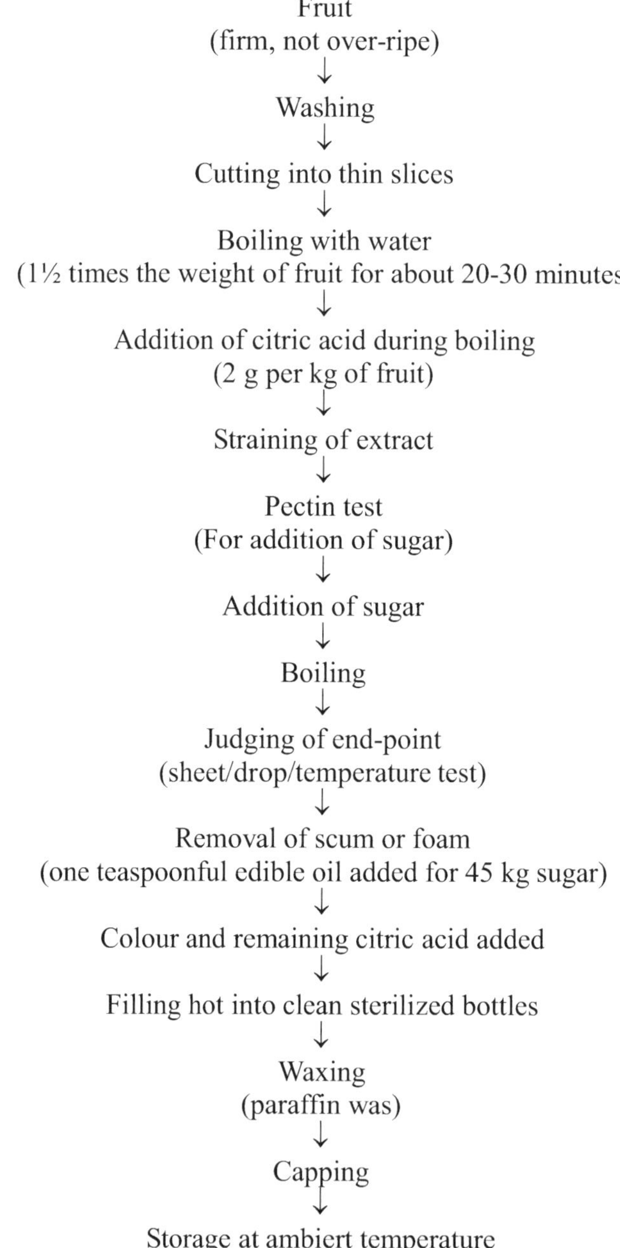

Figure 24: Flow diagram for processing of Jelly

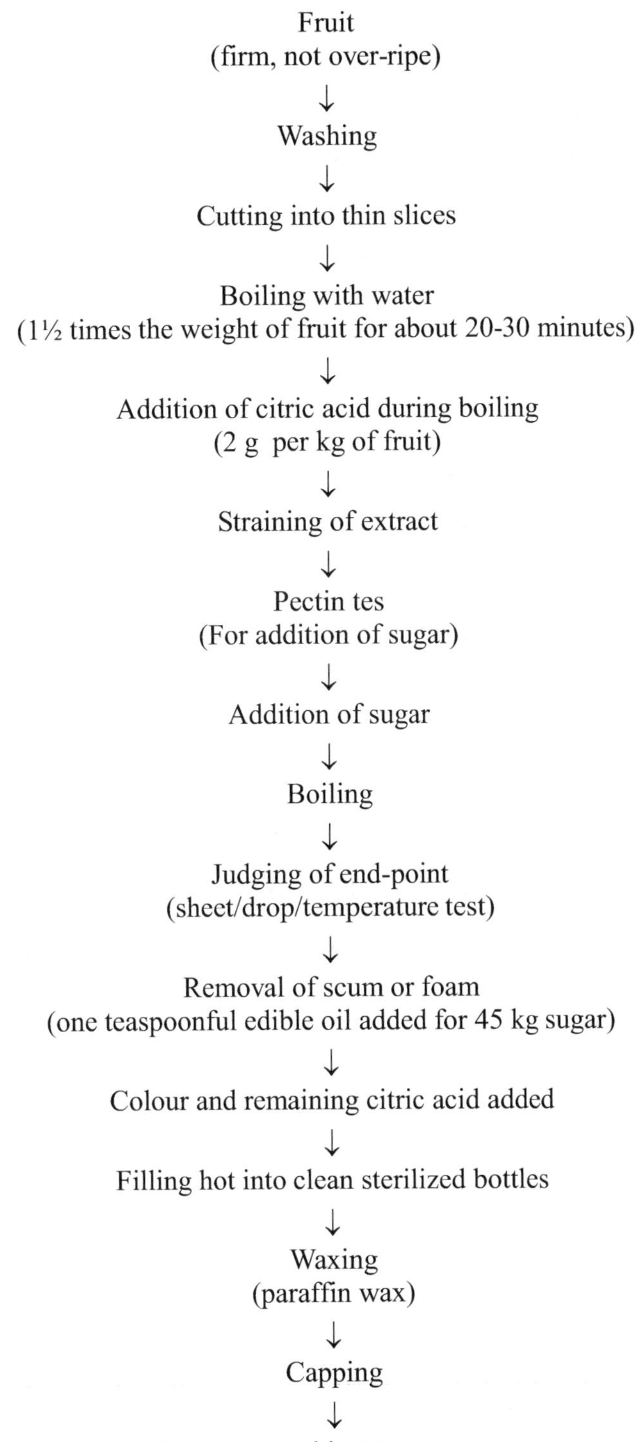

Figure 25: Flow diagram for processing of Marmalade

Crystallized Candy

Candied fruits/ vegetables when covered or coated with crystals of sugar, either by rolling in finely powdered sugar or by allowing sugar crystals to deposit on them from dense syrup are called crystallized fruits. The candied fruits are placed on a wire mesh tray which is placed in a deep vessel. Cooled syrup (70 per cent total soluble solids) is gently poured over the fruit so as to cover it entirely. The whole mass is left undisturbed for 12 to 18 hours during which a thin coating of crystallized sugar is formed. The tray is then taken out carefully from the vessel and the surplus syrup drained off. The fruits are then placed in a single layer on wire mesh trays and dried at room temperature or at about 49°C in driers.

IV. Preserve

A mature fruit / vegetable or its pieces impregnated with heavy sugar syrup till it becomes tender and transparent is known as a preserve. Aonla, bael, apple, pear, mango, cherry, karonda, strawberry, pineapple, papaya, etc. can be used for making preserves.

Intermediate-moisture foods or semi moist foods, in one form or another, have been important items of diet for a very long time. Generally, they contain moderate levels of moisture, of the order of 20-50% by weight, which is less than is normally present in natural fruits and vegetables, but more than is left in conventionally dehydrated products. In addition, intermediate-moisture foods contain sufficient dissolved solutes to decrease water activity below that required to support microbial growth. As a consequence, intermediate-moisture foods do not require refrigeration to prevent microbial deterioration. There are various kinds of intermediate-moisture foods : natural products such as honey; manufactured confectionery product high in sugar, jellies, jams, and bakery items such as fruit cakes; and partially dried products including figs, dates, etc. In all of these products, preservation is partially from high osmotic pressure associated with the high concentration of solutes; in some, additional preservative effect is contributed by salt, acid and other specific solutes.

Alcoholic Drinks

The most common examples of alcoholic beverages are wines and beers. Beer is usually made from a cereal, whereas wine can be produced from either cereals or fruit. Both can be distilled to produce spirits with an alcohol content of 30-50 per cent.

Both wines and beers are produced by fermentation which involves the conversion of sugars in the raw material or added sugar into alcohol and carbon dioxide. Different varieties of the yeast *Saccharomyces cerevisiae* are used to produce wines or beer. A simplified list of the differences is shown in the table opposite:

Table 21: Type of alcoholic beverages

Product	Type of yeast
Beer	'Top yeast' - *Saccharomyces cerevisiae*
Lager	'Bottom yeast' - *Saccharomyces carlsbergensis*
Fruit wine	'Wine yeast' - *S. oriformis, S. chevalieri,S. cerevisiae* (variety *ellipsoideus*) or a mixture of these
Palm wine	'Wild yeast'- natural mixture of yeasts
Rice wine	*Saccharomyces sake*

Although it is possible to use any strain of brewer's yeast for fermentation, it is necessary for a small producer to select one that works well and then continue to use it to produce a consistent product.

Alcohol has a lower boiling-point than water and distillation (vaporizing the alcohol and then condensing it) is used to concentrate the alcohol in spirit drinks. Distillation is carried out in stills which can be purchased for production at all levels. Alternatively, it is possible to construct basic still using locally-available materials (see below)

Figure 26: Diagram of Traditional still

A well-cleaned oil drum is fitted with a pipe to carry away the vapour, and a safety pipe. Alcoholic liquor is placed inside the drum and heated. On vaporization, the alcohol vapour is carried out of the drum via the pipe and passed through cooled air or cool water. The distillate condenses and is collected. Distillation is more frequently carried out on a centralized commercial level, although it does occur on a small scale.

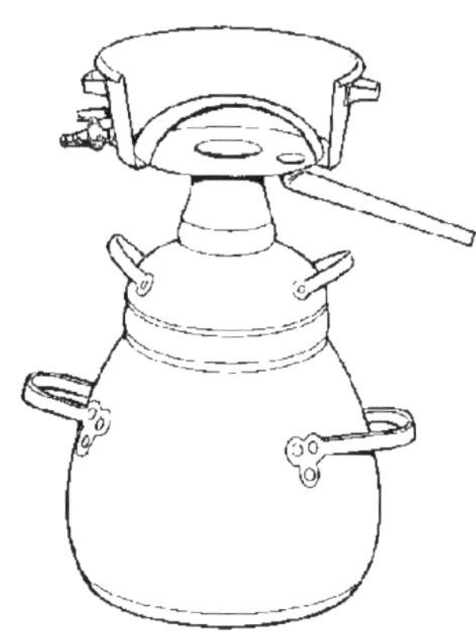

Figure 27: Diagram of Small still

Table: 22: Process and stages of beverage preparation

	Juice extraction/ pulping	*Mashing*	*Mix*	*Heat*	*Ferment*	*Filter*	*Bottle*	*Carbonate*	*Pasteurize*
Beer		*	*	*	*	*	*	*optional	*
Wine	*		*	*	*	*	*		
Sparkling wine	*		*	*	*	*	*		
Fruit juice	*					*	*		*
Carbonated drink			*				*	*	*

The table above outlines the processing steps for the production of a range of representative beverages.

Table 23: Equipment required for beverage processing

Processing stage	*Equipment*
Juice pulping/extraction	Fruit press /Pulper/Juicer
Mashing	Fermentation bins
Mixing	Mixers
Boil	Boiling pans
Fermentation	Fermentation bins/jars
Filter	Filters and filter presses
	Sieves
	Strainers
Carbonation	Carbonating equipment
Filling into bottles	Liquid fillers Funnel
Pasteurize	Open boiling pan
	Steam jacketed pan
	Pasteurizer

Beverage Processing

Pulping/Juice Extraction

Either the juice or the pulp from fruit is the starting material for the manufacture of soft drinks and wines.

Pulping

Soft fruits, such as papaya, can easily be pulped by hand or by using a pestle and mortar. A wide range of hand-operated pulpers are available, or if electric power is available, multi-purpose kitchen-scale equipment such as blenders can be used. At an industrial level, this process is normally carried out in pulpers which brush the fruit through a sieve and eject the skin and stones. Smaller models of this machine can be manufactured and are commercially available.

Extraction

Juice can be extracted from fruit in several ways.

- With a fruit press, fruit mill or hand pulper/sieve.
- By crushing/pulping with a mortar and pestle and then sieving through muslin cloth or plastic sieves.
- By steaming the fruit.
- Citrus fruit juices need to be extracted by reaming (squeezing) the fruit, and once again, comparatively simple equipment is available for this purpose.

Fermentation

As mentioned previously the process for achieving fermentation differs considerably depending upon the product. The following paragraphs describe the basic processes for the production of both beer and wine.

Beer

The process for making beer is often referred to as brewing. Brewing actually consists of three stages - mashing, boiling and fermentation.

Mashing involves the use of hot water (approximately 68°C) to extract the soluble materials from the malted grains. This produces a liquid called wort. The process is carried out in large vessels which may be made of wood or stainless steel.

The wort is then subjected to a process of boiling. In Europe this process involves the addition of hops. Boiling takes place in a similar vessel to the tubs used for mashing except that it is flask-shaped, with the neck being elongated in order to carry away the steam, and to prevent over-boiling.

Prior to inoculation (addition of the yeast), the wort is cooled. This is because if added to the hot wort the yeast would be inactivated. The degree to which the wort is cooled differs according to the type of beer to be produced. For example, fermentation for lager is conducted at 12-15°C using the yeast *Saccharomyces carlsbergensis*. In other cases, the yeast *Saccharomyces cerevisiae* is used at a temperature of 20°C. During fermentation, the beer is held in fermentation vats or food-grade plastic fermentation bins. When fermentation is complete, the process of packaging will depend on whether the beer is to be sold in draught form (e.g. in a keg), or if it is to be bottled and corked. If it is to be draught, the beer is not filtered and small amounts of yeast are left in it in order to keep it slightly carbonated. In the case of bottled beer, it is filtered and pasteurized.

Pulping Juice

Wine

In wine-making, the fruit juice or pulp is mixed with yeast and sugar and held in a fermentation bin. Again this may be made from food-grade plastic. This is left for about ten days during the first fermentation stage. Within 48 hours, fermentation becomes vigorous and there is frothing and foaming. It is important to keep the fermentation vessel closed to prevent bacteria and fungi from infecting the wine. After ten days the fermenting wine is racked.

This is done by scooping it up together with the solids using a sterilized mug, cup, or jug, and passing it through a muslin or nylon straining cloth. The cloth should have been sterilized and rinsed beforehand, and placed in a funnel. The wine is transferred into narrow necked fermentation vessels. These may be plugged with wads of cotton wool, or specially-designed vessels fitted with a airlock (known as a demijohn) may be used.

Ideally, fermentation is then continued at a temperature of 18°C. The whole process can take from three weeks to three months. The end of fermentation can be judged when it is seen that there are no more bubbles rising to the surface. At this stage, the wine is filtered, in order to remove the sediment from the wine and then syphoned into narrow-necked or food-grade plastic vessels, and stored for the minimum period in the recipe to allow the wine time to clear and mature before bottling. After this period of maturation, the wine is siphoned off into bottles and sealed with a sterilized cork-stopper or screw cap.

Carbonation

This involves the addition of carbon dioxide into a drink. The most usual way of achieving this is to use a pressurized cylinder or tank which contains a mixture of water and carbon dioxide. In the case of soft drinks, the bottle is filled to a certain level with the flavoured syrup, the bottle is positioned under the cylinder head and carbon dioxide is released. The bottles are capped immediately. Cylinders for holding carbon dioxide are available for both large-scale production and in smaller sizes for use at the household level.

Pasteurization

Liquid products such as drinks may need to be pasteurized if they are to have a shelf-life of more than a few days. Pasteurization involves heating the product to a temperature of 80-90°C and holding it at that temperature for between 0.5 and 5 minutes before filling into clean sterilized bottles. Pasteurization is best carried out over a direct heat in stainless steel pans.

Some products can be pasteurized in their bottles. The filled bottles, with the lids loosely closed, are stood in a large pan of boiling water with the water-level around the shoulder of the bottle.The time and temperature required for pasteurization will depend on the product and the bottle size.

Packaging

Beverages have differing needs with regard to storage, but the most pressing need for all beverages is simply that they need to be contained without the possibility of leakage. The tables below outline some of the other storage requirements and the suitability of different types of container.

Table 24: Different type of storage requirements

	Light	*Air*	*Heat*	*Micro-organisms*	*Insects*
Fruit juice, cordial etc.	some	*		*	*
Beer	*	*	*	*	*
Wine	*	*		*	*
Soft drinks				*	*

Glass bottles are the most popular medium for packaging beverages. However, owing to the expense of new glass, many producers (particularly those operating on a small scale) re-use the bottles. This means that in order to prevent contamination the bottles must be sterilized and cleaned properly. Simple hand-held bottle-brushes can be used to ensure a good standard of cleanliness, and mechanized brush-cleaners are also available.

Table 25: Type of package in different containers

	Glass bottle/jar	*Metal can*	*Plastic film/pot/pouch*	*Ceramic pot*
Fruit juice cordials etc.	*	Lacquered	*	*
Beer	Coloured	Lacquered	*	*
Wine	*			*
Soft drinks	*	*		

Most beverages are thin liquids and can be filled quite easily by hand, but this is often too slow for a small business. A simple filler can be made by fitting one or more taps to the base of a bucket (see diagram below).

The bucket should be made from stainless steel for hot acid liquids (e.g. fruit juices) or food-grade plastic for cold filling. Iron and copper should not generally be used in food handling.

The type of closures used depends upon the type of product and its particular use (e.g. for glass bottles does it need to withstand internal pressure from carbonation).

Figure 28: Diagram for Liquid filler

There is a large range of closures available for glass bottles, but the choice for small-scale producers may often be restricted by what is locally available.

Metal 'crown' caps are commonly used for beers and fruit juices, whereas squashes, carbonated drinks and spirits are more frequently packaged using re-sealable metal screw-caps. Wine is often sealed with a cork but plastic stoppers are equally effective and cost less.

With technological advances in the field of packaging materials, larger commercial manufacturers are using formed waxed cartons for beverages such as fruit juice. These have become very popular since they are cheaper and more convenient. Unfortunately the cost of the equipment needed to form and seal the cartons is very expensive and is presently out of reach for the small-scale producer.

Cheaper alternatives include plastic or foil laminated pouches. If sealed correctly, they can be a very convenient way of packaging.

Beverages can also be canned, but the cost of aluminium or steel cans is usually prohibitive for small-scale producers. In addition, the correct type of lacquer is required on the inside of the cans, and they are not re-usable.

V. Dehydration Process

Dehydration means the process of removal of moisture by the application of artificial heat under controlled conditions of temperature, humidity and air low. In this process a single layer of fruits, whole or cut into pieces or slices are spread on trays which are placed inside the dehydrator. The initial temperature of the dehydrator is usually 43°C which is gradually increased to 66-71°C for fruits.

Basic types of drying process

- Sun drying and solar drying
- Atmospheric drying including batch (kiln, tower and cabinet driers) and continuous (tunnel, belt, belt-trough, fluidized bed, puff, foam-mat, spray, drum and microwave);
- Sub-atmospheric dehydration (vacuum shelf/belt and freeze driers).

Sun and solar drying of fruits and vegetables is a cheap method of preservation because it uses the natural resource / source of heat; sunlight. This method can be used on a commercial scale as well as the village level provided that the climate is hot, relatively dry and free of rainfall during and immediately after the normal harvesting period.

- Large scale driers are more promising than small scale ones. However, small scale driers should not be neglected.
- The drier should be designed to maximize the utilization factor of the capital investment, i.e. multi-products (fruits, vegetables and other raw material) and multi-use (eg. Drying and heating water for domestic use).
- In general, an auxiliary heat source should be provided to assure reliability, to handle peak loads and also to provide continuous drying during periods of no sunshine.

Shade Drying

Shade drying is carried out for products which can loose their colour and / or turn brown if put in direct sunlight. Therefore, shade drying is carried out under a roof or thatch which has open sides.

Osmotic Dehydration

In osmotic dehydration the prepared fresh material is soaked in a heavy (thick liquid sugar solution) and / or a strong salt solution and then the material is sun or solar dried.

Common driers used for drying / dehydration

A. Air Convection Driers

- Kiln drier
- Cabinet, tray and pan driers
- Tunnel and continuous belt driers
- Belt trough drier
- Air lift drier
- Fluidized bed drier
- Spray driers

B. Drum or Roller Driers

C. Vacuum Driers

- Vacuum shelf driers
- Continuous vacuum belt drier
- Freeze-drying

VI. Pickle Production

The preservation of food in common salt or in vinegar is known as pickling. It is one of the most ancient methods of preserving fruits and vegetables. Pickles are good appetizers and add to the palatability of a meal. They stimulate the flow of gastric juice and thus help in digestion. At present, pickles are prepared with salt, vinegar, oil or with a mixture of salt, oil, spices and vinegar.

Fruit based beverages are relished when served chilled, particularly during summers. These are delicious as well as nutritious containing the goodness of fresh fruit. Squash, syrup, cordial, crush are diluted with water before use. The juice is used as such. Ready-to-serve beverages are made out of juice, sugar and water and consumed as such. The fruits e.g. pineapple, orange, lime, banana, litchi, passion fruit, other local fruits can be used. The final products are pure juice, ready-to-serve beverages packed in 200 ml glass bottles; squashes/syrups packed in 500 ml/700 ml/ 1L glass or PET bottles.

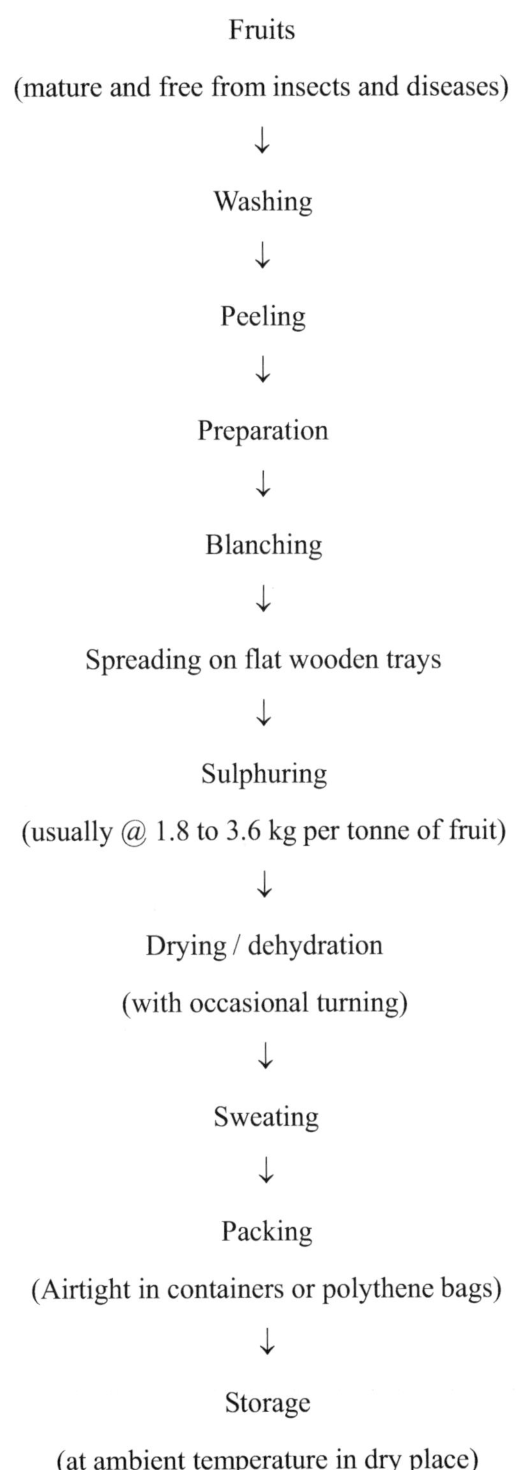

Figure 29: Flow diagram for Drying/dehydration fruits/vegetables

References

Addy, N.D. and Stuart, D.A. 1986. Impact of biotechnology on vegetable processing. Food Technol. 40(10), 64-66

Andrews, S., Food and Beverage Science , Tata McGraw Hill, Publishing Co. Ltd., New Delhi, 1980

Arthey, D. and Ashurst, P. 1995, Fruit Processing, Chapman & Hall, London, New York.

Arthey, D. and Dennis C, P. 1991, Vegetable Processing, Chapman & Hall, London, New York.

Green.L.F.1978. Developments in Soft Drinks Technology. Applied Science Publishers, London.

Hicks, D.1990. Production and Packaging of Non- Carbonated fruits Juices and Fruit Beverages. Chapman & Hall, London, New York.

Hunt, P. (ed). Fruit and Vegetables, Ward Lock Ltd., London, 1972.

Jasscler, D.K. and M.A. Joslyn (eds) Fruit and Vegetables Juice, The AVI Publishing Co. Inc., Connecticut, 1971.

Jagtiant, J. 1988. Tropical Fruit Processing. Academic Press, San Diego, CA.

Lal, G, G.S. Siddappa and G.L. Tandon, Preservation of Fruits and Vegetables, Indian Council of Agricultural Research, New Delhi, 1960.

Mitchell, A.J.1990. Formulation and Production of Carbonated Soft Drinks. Chapman & Hall, London, New York

Nelson, P.E and Tressler, D.K. 1980, Fruit and Vegetable Juice Processing Technology. AVI Publishing Co., Wesport, CT

Nagy, S., P.E. Shaw, and M.K. Veldhnis, Citrus Science and Technology, The AVI Publishing Co.Inc., Connecticut, 1977.

Nath, P. Vegetables for the Tropical Region, Indian Council of Agricultural Research, New Delhi, 1977.

Salunkhe, D.K., Bolin, H.R., and Reddy,N.R.1990. Storage, Processing , and Nutritional Quality of Fruits and Vegetables.2nd ed. CRC Press, Boca Raton, FL.

Varnam, A.H. 1994.Beverages: Technology, chemistry and Microbiology. Chapman & Hall, London, New York.

Vine, R.P.1981. Commercial Winemaking, processing and controls. Chapman & Hall, London, New York.

Woodroff, J.G. and Phillips, G.F. 1981. Beverages Carbonated and Noncarbonated .rev. ed. AVI Publishing Co., Westport, CT.

Chapter - 4

Tropical Root and Tuber Crops Based Foods

Introduction

The tropical root and tuber crops are comprised of crops covering several genera. They are staple foods in many parts of the tropics, being the source of most of the daily carbohydrate intake for large populations. These carbohydrates are mostly starches found in storage organs, which may be enlarged roots, corms, rhizomes, or tubers. Many root and tuber crops are grown as traditional foods or are adapted to unique ecosystems and are of little importance to world food production. Others such as cassava (*Manihot esculenta* Crantz) and white-fleshed sweet potato (*Ipomoea batatas* L.) are known worldwide. Several of these crops have been termed under-exploited and deserving of considerably more research input

Potato and Banana chips: Potato chips are one of the popular snack foods consumed throughout the country. Chips are mostly prepared by small scale artisans / confectioners, and sold in plastic film or paper packs. Recently, however some of the companies have installed plants having relatively large capacities and have started marketing chips in attractive packs. The fat content generally ranges between 25-40% and moisture content between 2-4%. Potatoes having high total solids, low reducing sugars and low amylase activity are most suitable for chips manufacture.

Banana chips are prepared from varieties rich in starch having very low levels of soluble sugars and polyphenol - oxidase activity. These are mostly processed and marketed in Southern States especially in Kerala. Previously, the chips were fried in coconut oil, but now other vegetable oils are also used in their processing. Relatively, banana chips are harder in texture than potato chips.

Cassava

The presence of hydrocyanic glucosides (HCN) in all plant parts presents some problems in marketing cassava. Selections have been made from both chance seedlings and in breeding programs which are low in HCN. These are the only types sold in U.S. markets. More than one kg of unprocessed roots would have to be consumed before lethal doses of HCN would be reached. Peeling and boiling in water are common methods of removing a large proportion of the HCN in the roots. Other postharvest problems with cassava include proper handling and storage of cuttings under frost-free conditions.

Roots are usually peeled and boiled or baked. Commercial processing of cassava is limited to packers of frozen, peeled roots, which are marketed in India. in packages like frozen french fries. This convenience pack may have the potential for expanded utilization of cassava in this country and Europe. Deep-fried chips, like potato chips, are produced and marketed in many states of India. Cassava starch, known as tapioca, has limited potential for expansion. Even though cassava flour can be used as a partial substitute for wheat flour in the production of bread,

Sweet Potato

Sweet potato is one of the world's important food crops and an important staple food crop many countries. It is very valuable in the diet of rural poor in the tropics. It is a low-input crop and it is used as a vegetable, a dessert, source of starch and animal feed. In India, sweet potato is eaten as a substitute for yam as a result of lower cost of production. Sweet potato is an important food crops. The sweet potatoes are either baked or boiled.

Sweet potato is a food security crop. It is drought-resistant, provides good ground cover, and grows on soils with limited fertility. The roots and leaves are sources of carbohydrates, proteins and minerals. They are eaten as a vegetable after boiling, baking, or frying and sometimes sliced and sun-dried to produce chips, which are ground into sweet potato flour.

Yams

Several types of yams are grown in the tropics and subtropics. Some, which will not be discussed here, are grown only for medicinal purposes. Of the edible species, *Dioscorea alata* L., known as the Greater Yam, *D. cayenensis* Lam., the Yellow Yam, and *D. rotundata* Poir., the White Yam, are the most common. Tuber flesh varies from white to yellow and is from 15 to 40% starch. Tubers have a distinct dormancy period, which can be extended with curing and the application of gibberellic acid. This makes yams ideal for long distance shipment and export. Yams are usually baked or boiled and mashed. Unless the production expenses are reduced, little potential exists for commercial processing of yams into items like potato chips or French fries.

Edible Aroids

Corms which contain 25 to 35% starch are plagued by the presence of an acrid factor, which causes itchiness and considerable inflammation of tissues. Cooking removes most, if not all, of this factor from domesticated clones. Shelf life of harvested corms varies considerably between taro and cocoyam and depends on the care taken during the harvesting and packaging process. Cocoyam has a considerably longer shelf life of several weeks. This can be extended further with curing and refrigerated storage. Corms are usually peeled and boiled. Processing is limited to the production of deep-fried chips.

Cassava Flour and Starch

Separation of the starch granules from the tuber in as pure a form as possible is essential in the manufacture of cassava flour. The granules are locked in cells together with all the other constituents of the protoplasm (proteins, soluble carbohydrates, fats and so on), which can only be removed by a purification process in the watery phase. Processing the starch can therefore be divided into the following stages:

1. Preparation and extraction. Crushing of the cells and separation of the granules from other insoluble matter (i.e.' adhering dirt and cell-wall material) including the preparatory operations of washing and peeling the roots, rasping them and straining the pulp with the addition of water.
2. Purification. Substitution of pure water for the aqueous solution surrounding the starch granules in the mash obtained in the first stage, as well as the operations of sedimentation and the washing of the starch in tanks and on flour tables, silting, centrifuging, etc.
3. Removal of water by centrifuging and drying.
4. Finishing. Grinding, bolting and other finishing operations.

This method of processing is essential in the preparation of any kind of starch. For cassava, however, because of the relatively small amount of secondary substances, the separation at each stage is performed with great ease. Whereas with maize and other cereals the grinding of the seed and the mechanical separation of the germ and the pericarp from the grain present special problems in stage 1, and the separation of protein and other constituents in stage 2 can only be accomplished with the aid of chemicals, these operations can be reduced to a minimum in cassava preparation. It is indeed possible to obtain first-rate flour from the cassava root without special equipment by using only pure water. This makes the processing of cassava flour particularly suitable for rural industries.

Processing Operations

Importance of Quick Processing

In the processing of cassava starch it is vital to complete the whole process within the shortest time possible, since as soon as the roots have been dug up, as well as during each of the subsequent stages of manufacture, enzymatic processes are apt to develop with a deteriorating effect on the quality of the end product. This calls for a well-organized supply of roots within relatively short distances of the processing plant and, furthermore, for an organization of the stages of processing that will minimize delays in manufacture. Thus, while simple in principle, the manufacture of good cassava flour requires great care.

The roots are normally received from the field as soon as possible after harvest and cannot be stored for more than two days. Since the presence of woody matter or stones may seriously interfere with the rasping process by stoppage or by breaking the blades, the woody ends of the roots are chopped off with sharp knives before the subsequent processing operations.

Peeling and Washing

In small and medium-size mills the general practice is to remove the peel (skin and cortex) and to process only the central part of the root, which is of much softer texture. With the relatively primitive apparatus available and limited power, the processing of the whole root would entail difficulties in rasping and in removing dirt, crude fiber and cork particles, whereas comparatively little extra starch would be gained.

The structure of the root permits peeling to proceed smoothly by hand (it is often done by women and children). Work starts in the morning as soon as the roots are brought in; as it must be finished as quickly as possible, numerous hands are needed. The roots are cut longitudinally and transversely to a depth corresponding to the thickness of the peel, which can then be easily removed. Any dirt remaining on the smooth surface of the core of the root can now be washed off without any trouble and the peeled roots deposited in cement basins where they remain immersed in river water until taken out for rasping. Frequent treading by foot cleans any loosely adhering dirt from the roots.

In the larger factories, whole roots are generally processed. The washing here serves to remove the outer skin of the root as well as the adhering dirt. Provided the root is sufficiently ripe, skin removal may proceed without the use of brushes. Only the outer skin or corky layer is removed, as it is profitable to recover the starch from the cortex.

The mechanical washer is a perforated cylindrical tank which is immersed in water. A spiral brush propels the roots while they are subjected to vigorous scrubbing in order to remove all dirt. A centrifugal pump is fitted to one end of the machine and connected to a series of jets arranged along the carrying side of the brush. These jets produce a countercurrent to the flow of the roots, ensuring that they receive an efficient washing.

Another efficient washer is a rotary drum with an interior pipe which sprays water on the roots. The drum is either wooden or perforated metal, about 3 to 4 m long and 1 m in diameter, with horizontal openings; it is mounted inside a concrete tank. In some, rotating paddles are fitted along the axis. Washing is done by the action of water sprayed, assisted by the abrasion of the roots both against one another and against the sides of the cylinder or the paddles.

The roots are hand-fed from one end and when they come out at the other they are clean and partially peeled, the action being continuous. Dirty water and skin are periodically drained out through a small opening in the concrete tank. In modern factories the roots are pre-washed by soaking in water to separate the coarse dirt and then passed through a combined unit for washing and peeling as described above.

Rasping or Pulping

It is necessary to rupture all cell walls in order to release the starch granules. This can be done by biochemical or mechanical action. The biochemical method, an old one, allows the roots to ferment to a certain stage; then they are pounded to a pulp and the starch is washed from the pulp with water. This method does not give complete yields and the quality of the resulting starch is inferior. Mechanical action is carried out by slicing the roots and then rasping, grating or crushing them, which tears the flesh into a fine pulp.

By pressing the roots against a swiftly moving surface provided with sharp protrusions, the cell walls are torn up and the whole of the root is turned into a mass in which the greater part, but not all, of the starch granules is released. The percentage of starch set free is called the rasping effect. Its value after one rasping may vary between 70 and 90 percent: the efficiency of the rasping operation therefore determines to a large extent the overall yield of starch in the processing. It is difficult to remove all the starch, even with efficient rasping devices, in a single operation. Therefore, the pulp is sometimes subjected to a second rasping process after screening. The rasping is carried out in different ways with varying efficiency.

Hand and Mechanical Rasping

On very small holdings in some cassava-growing regions the roots are still rasped by hand on bamboo mats. Where daily production amounts to several hundred kilograms of flour, simple mechanical implements are used.

A simple but effective grater is obtained by perforating a sheet of galvanized iron with a nail and then clamping it around a wheel with the sharp protruding rims of the nail openings turned outward. The wheel may be driven by hand, but it is often driven by foot like a tricycle, the worker pressing the roots from above onto the rasping surface: or the rasping surface is attached to one side of a rotating disk equipped with a crank transmission, which is driven by foot. The pulp is collected in baskets or wooden containers to be carried to the sieves.

Hydraulic Raspers

Larger water-powered raspers can be used where running water is available. The waterwheel is rotated by a flywheel and driving belts to a pulley on the shaft of the rasping drum. The drum 20-30 cm in diameter is either attached to a primitive wooden construction or fitted into a "rasping table" The operator seated at the table, presses the roots against the drum. The grated mass is forced through a narrow slit between the drum and the shelf before it drops into the trough, whence it is carried to the sieves. The rasping devices described above are made of perforated in plate. Though inexpensive, they are relatively inefficient as the rasping plate must often be replaced on account of rapid wear.

Engine-Driven Raspers

Engine-driven raspers are more economical when production rises above a certain level - say, for the handling of 10 tons of fresh roots a day. The machine has a rotor of hardwood or drawn steel tube, 50 cm in diameter, with a number of grooves milled longitudinally to take the rasping blades or saws. The number of saw teeth on the blades varies from 10 to 12 per centimeter according to need. The blades are spaced 6-7 mm apart on the rotor.

In simpler versions, the rotor is fitted into housing in such a way that the rasping surface forms part of the back wall of the receptacle for the roots. Facing the rasping surface, a block or board is inserted which is movable by a lever and turns on an axis near the upper rim of the compartment. By manipulating this buffer the roots are pressed onto the rasping surface, which moves downward in the hopper, and the mass is propelled through a slit in the bottom of the hopper. It is advisable to give the inner surface of the buffer the form of a circular segment corresponding to the section of the rotor exposed so that, at its extreme position inward, the distance between rotor and block is only a few millimeters. This, however, is generally possible only in the all-steel raspers to be described later.

In a rasper of the type used in larger factories, the housing is equipped with adjustable breasts with sharp steel edges for the control of rasping fineness. More recent constructions provide for the return to the rasping surfaces of those pieces of the roots which were thrown out sideways. The pulp has to pass a screen-plate with sharp-edged holes or slits, during which it is homogenized to a certain degree and, in fact, undergoes a secondary crushing.

Power Consumption During Rasping

To obtain the maximum rasping effect, the power supply should be accurately attuned to the constructional details of the rasper, i.e., to the distance between the surface of the rotor and the breasts in the housing. The energy required to tear up the roots is derived from the momentum of the rotor, a certain minimum of kinetic energy being necessary to obtain any rasping effect. Above a certain rotor speed, however, it is to be expected that no considerable further increase in the rasping effect will be obtained. There is thus an optimum speed for the rasper in conformity with the need for a high rasping effect on the one hand and with the economy of power supply on the other. In this connection it should be remembered that only the linear velocity of the rasping surface counts. In practice it has been found that a rasper of the usual dimensions - a diameter of 40-50 cm and a length of 30-50 cm for the rotor - should be driven at 1000 rotations per minute, corresponding to a linear velocity of the rasping surface of about 25 m per second. The power of the engine required to drive a single rasper of this type is 20-30 hp. In most cases diesel engines are used.

Secondary Rasping or Grinding

In view of these results it is no wonder that the rasping effect differs widely in different factories. In modern factories, it may be estimated that an effect of about 85 percent is attained at the first rasping; at these production levels, however, it is economical to submit the pulp to a second crushing process, either in a second rasper or in special mills where the pulp is ground between stones. These mills, however, do not seem to have found much favour with cassava manufacturers.

In a secondary rasper, the indentation of the saw blades should be somewhat finer, about 10 per centimeter (25 to 27 teeth per inch) as compared with about 810 per centimeter (19 to 26 teeth per inch) for the primary rasper. The overall rasping effect is raised to over 90 percent by the secondary rasper.

Screening

In separating the pulp from the free starch a liberal amount of water must be added to the pulp as it is delivered by the rasper, and the resulting suspension stirred vigorously before screening. Mixing with water can be carried out more or less separately from screening, but more often the two operations are combined in "wet screening" - that is' the mass is rinsed with the excess water on a screen which is in continuous motion.

Hand Screening

In the smallest mills, screening is done by hand. The rasped root mass is put in batches on a cloth fastened on four poles and hanging like a bag above the drain leading directly to the sedimentation tanks. Spring water or purified river water is run in from a pipe above the bag, and the pulp is vigorously stirred with both hands. Sometimes bamboo basketwork is used to support the screening cloth. The pulp under processing still contains appreciable amounts of starch and therefore has a certain value (e.g., as a cattle fodder): in the small mills it is pressed out by hand, and the lumps obtained are dried on racks in a well-ventilated place.

The Rotating Screen

A simple form of rotating screen consists of a conical frame of hardwood, fixed on a hollow, horizontal axis, at least 3 m long, covered with ordinary cloth or phosphor-bronze gauze. Phosphor-bronze is often preferred for its durability, but its use necessitates frequent brushing in order to remove clogging pulp particles. The crude pulp is fed into the cone at the narrow end and by the rotation of the screen, at approximately 50 revolutions per minute, slowly moves down to the other end, whence it is conveyed to the pulp tanks. In the meantime, water is sprayed on it under pressure (e.g., 6 atmospheres) from a number of openings in the hollow shaft. Thus, by the time the pulp reaches the lower end of the cone, it is more or less completely washed out. The rotation screen has the advantage of preventing the plugging of the meshes of the sieve with gummy materials (they tend to agglutinate with the fibre as the screen rotates). The flour milk is caught in a cemented basin stretching out below the screen over its whole length, and from there runs along channels into sedimentation tanks or flour tables.

The Shaking Screen

In large factories the rotating screen is replaced by the shaking screen. It consists of a slightly inclined, horizontal frame, 4 m in length and covered with gauze, which is put into a lengthwise shaking motion in short strokes by means of an eccentric rod. The fresh pulp, after being mixed with water in distribution tanks, is conducted by pipes to the higher end of the screen; during screening, the pulp remaining on top of the screen is slowly pushed downward by the shaking motion.

It is advantageous to let the suspensions pass a series of shaking screens of increasing fineness (80-, 150-, and 260-mesh), the first one retaining the coarse pulp, the others the fine particles. The pulp remaining on the first of these screens is often subjected to a second rasping or milling operation and then returned to the screening station.

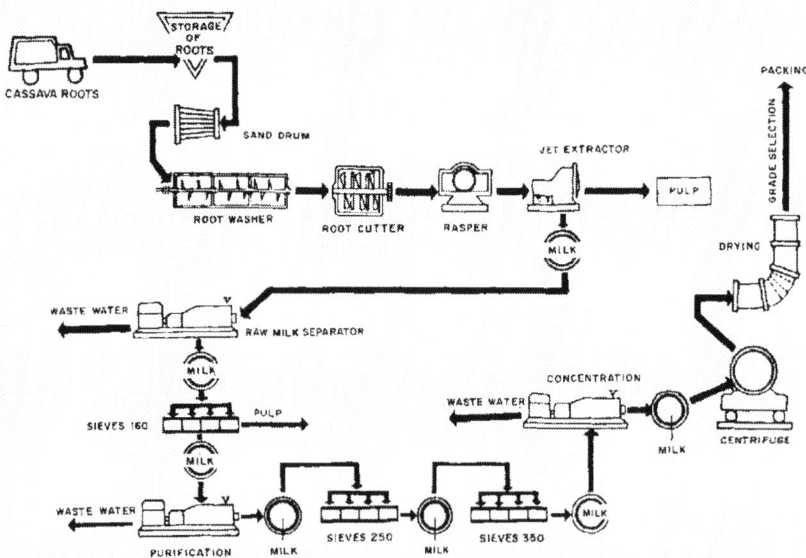

Figure 30: Diagram for Medium scale cassava processing unit

Extraction of Starch from Dried Cassava Root

A limited quantity of the cassava imported into Europe in the form of chips and dried sliced roots is manufactured into starch. The dried roots are cleaned, washed and grated and the starch is separated by cylindrical sieves; however, this practice is costly and the starch produced is of inferior quality for the following reasons.

1. The brown skin, which contains chlorophyll and coagulated proteinous substances, adheres strongly to the ligneous tissues. While it is easy to remove this skin from the fresh roots, it is very difficult to remove it from the dried roots and, therefore, the starch of dried roots is always dark.
2. The nitrogenous substances are found in a colloidal state enveloping the starch granules in the pulp slurry of the fresh roots. It is easier to separate these nitrogenous particles in the pulp slurry of fresh roots than in dried roots.

Baked Tapioca Products

The baked products for which cassava flour is the basic ingredient are known commercially as tapiocas or tapioca fancies. In Malaysia and some other areas these products are commonly known in the industry as sago products. The term probably originated with the Chinese production of sago-palm starch products. The manufacture of tapioca fancies is a logical follow-up of the production of the flour itself in the countries of origin. Separation of the processing of the flour and of the derivatives would be illogical. Many medium-size and larger factories are also equipped for the manufacture of such baked products as flakes, seeds, pearls, and grist. These products are made from partly gelatinized cassava starch obtained by heat treatment of the moist flour in shallow pans. When heated, the wet granules gelatinize, burst, and stick together. The mass is stirred to prevent scorching. They are manufactured in the form of irregular lumps called flakes or of perfectly round beads 16 mm in diameter known as seeds and pearls The grist is a finer-grained product obtained by milling gelatinized lumps, and siftings and dust are residual products of the manufacture of seeds and pearls.

Preparation of Wet Flour

The raw material for baked products is the flour scooped up from sedimentation tanks or tables after the supernatant or excess water, has been drained and the "yellow" flour scraped off. Clearly the use of moist starch, an intermediate stage in the processing of the Flour, is economically advantageous.

Only very white first-quality flour can be used in the manufacture. To obtain this, sulfurous acid is often added in the first sedimentation. This chemical should, (however) be washed out as completely as possible by a second sedimentation in clean water; any traces of the acid left in the flour tend to spoil the quality of the end product. It is strongly advised not to use active chlorine preparations in this case, as they influence the agglomeration of the starch into pearls and other forms in an unfavourable way.

The cake of moist flour, containing about 45 percent water, is broken up by a small mill, spades or pressing it through frames strung with steel wire spaced about 10-20 cm apart, after which the lumps are rubbed through a screen of about 20 mesh/inch to produce a coarse-grained moist flour.

At this stage the flour is ready only for gelatinization and the production of flakes; to prepare pearls and seeds, the small aggregates of moist starch should be subjected to a process of building up and consolidation. This gives them the size and cohesive strength desired for the further treatment. The operation is known by the Indonesian name as the gangsor method. A portion of the moist starch is put into a long cylindrical bag of twill cloth which is held at each end by one man. Together with a rhythmical strong jerking movement, they throw the mass of starch lumps from one end of the hag to the other After a few minutes of this treatment the irregular lumps have grown into beads of varying size and have gained in firmness. Another portion of the moist flour is added and the gangsoring is continued the operation being repeated until the heads have grown more or less to the desired size. Depending on the skill of the worker the size of the starch balls is fairly uniform. Curiously enough, the knack of gangsoring is achieved only by a fraction of all workers, so the operation should be classified as skilled labour.

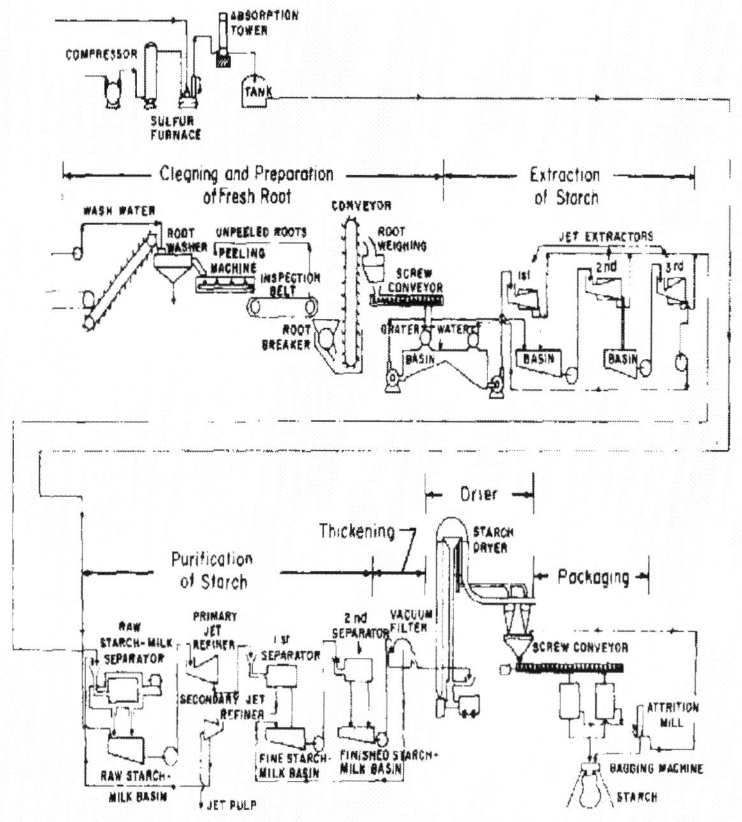

Figure 31: Diagram for Modern cassava processing plant

In Malaysia the flour is fed into open, cylindrical rotating pans about 0.9 m in diameter and 1.2 m deep. During rotation the starch grains are forced to adhere together in the form of small particles or beads. The resulting product depends on the speed and the length of time of rotation. After gangsoring, beads of the right size are sorted out by screening between plates with circular holes corresponding to the required dimensions.

Gelatinization

In gelatinizing, starch undergoes a radical alteration in molecular arrangement, with a concomitant change in properties. From a practically insoluble product of semicrystalline structure it becomes an amorphous substance, miscible with water in any proportions at sufficiently high temperatures, giving viscous solutions which after cooling set to a semisolid elastic mass: a jelly, or gel.

This process may be brought about by the action of chemicals or by heating in an aqueous medium; only the latter case is of interest here. The onset of gelatinization is characterized by a loss of granular, structure which also promotes swelling; both processes can easily be followed under a microscope. With cassava starch, gelatinization sets in at about 60°C, and the process is completed at about 80°C. The point of gelatinization depends to a certain extent on granule size, the smaller granules being more resistant to swelling.

In the manufacture of baked products, the treatment is kept at a moderate temperature so as to cause gelatinization only in the surface layer of the lumps of moist starch. The product obtained therefore consists of agglomerations of practically raw starch enclosed by a thin layer of the tough and coherent gelatinized form.

The hand-baking process can also be applied in the manufacture of pearls and seeds, but rather irregularly shaped beads are obtained, inferior in colour and in other qualities. Better mechanical methods for obtaining a first-rate product have long been known. In one of these, gelatinization is performed with the direct application of steam. The starch beads are poured onto plates in a rather thick layer, the plates forming a conveyor belt which is slowly drawn through a tunnel charged with steam. In this way, uniform gelatinization is ensured.

Drying

The gelatinization process in the hand-worked flakes changes the moisture content of the product by no more than a few percent, and the same applies to the steam-treated pearls and seeds. In the drum described above, drying sets in parallel with the gelatinization and may be promoted by ventilating the drum, but the removal of water here is also incomplete. Thus, in general, a final drying after gelatinization is necessary in order to bring down the moisture content to the desired level of about 12 percent. Drying in this case is best accomplished in chamber driers of the circulating type. For instance, in a chamber drier for pearls and seeds the initial temperature should not exceed 40°C lest further gelatinization and bursting of the beads set in. Toward the end of the treatment the temperature may be raised to 60-70ºC. With efficient exhausters, drying may be completed in 1 1/2-2 hours. Normally, from 16 tons of moist starch, 10 tons of the dried products are obtained

Processing Cassava Chips

The present method of processing chips in Thailand, Malaysia and some other countries is very simple, consisting in mechanically slicing the cassava roots and then sun drying the slices. The recovery rate of chips from roots is about 20-40 percent. However, the products are considered inferior in quality by some quality-conscious feedstuff manufacturers, although many others consider them satisfactory.

Preparation of the Roots

When the roots are not sorted, peeled and washed, the chips are usually brown in colour and have a high content of fiber sand and foreign objects as well as hydrocyanic acid. Trimming, peeling and washing the roots in a similar manner as for the processing of cassava flour are recommended in order to produce white chips of superior quality.

Slicing or Shredding

The roots are shredded in a special machine, which is usually made locally. The machine consists of a rotating notched cutting disk or knife blades mounted on a wooden frame equipped with a hopper. The cassava roots are cut into thin slices and pieces as they pass through the machine.

Drying

Sun drying is used mostly where the sliced roots are spread out on drying areas, or concrete floors of various dimensions. Experiments in Madagascar showed that the concentration of chips during drying should not exceed 10-15 kg/m2, the required drying area space being about 250 m2 for each ton per day of dried roots produced.

To produce good quality chips the roots must be sliced and dried as quickly as possible after harvest. The chips should be turned periodically in the drying period, usually two or three sunny days, until the moisture content reaches 13 to 15 percent. The chips are considered dry when they are easily broken but too hard to be crumbled by hand. The thickness of the slices also has an effect on the quality of chips. Thick slices may appear dry on the surface when their internal moisture content is still high.

When rain threatens during the drying process, the chips are collected by hand or by a tractor into piles under a small roof. Interrupted sun drying affects the quality of the finished chips and pellets. When the semidried chips are wet again by rain, they become soggy and upon completion of drying lose their firm texture. In rainy regions, where continuous sun drying is difficult, some form of artificial heat drying is required.

Broken Roots

Similar to chips in appearance, but generally thicker and longer, they are often 12-15 cm long and can jam the mechanism of handling equipment. They are produced mainly in Africa where local processors prefer to produce longer roots because of the domestic demand mainly for products suitable for human consumption, as cassava is part of the staple diet. Once processed into chips the product becomes inedible, and the producer wants to conserve the local market.

Pellets

The pellets are obtained from dried and broken roots by grinding and hardening into a cylindrical shape. The cylinders are about 2-3 cm long and about 0.4-0.8 cm in diameter and are uniform in appearance and texture. The production of pelleted chips has recently been increasing as they meet a ready demand on the European markets. They have the following advantages over chips: quality is more uniform; they occupy 25-30 percent less space than chips, thus reducing the cost of transport and storage; handling charges for loading and unloading are also cheaper; they usually reach their destination sound and undamaged, while a great part of a cargo of sliced chips is damaged in long-distance shipment because of sweating and heating.

Pellets are produced by feeding dried chips into the pelleting machine, after which they are screened and bagged for export. The powdered chips which fall down during pelleting are re-pressed into pellets and the process is repeated. There is usually about 2-3 percent loss of weight during the process.

Meal

This product is the powdered residue of the chips and roots after processing to extract edible starch. It is generally inferior in quality to chips, pellets and broken roots, has a lower starch content and usually contains more sand. The use of cassava meal in the European Economic Community has declined with a shift to the other cassava products during the last few years. However, there will remain some demand for this product, especially by small scale farmers who produce their own feedstuffs. Since it does not require grinding and thus can be readily mixed with other ingredients.

Modified Starches

The modification techniques, namely, chemical, physical and enzymatic modification, the chemical method is the most important changed in its physical and/or chemical properties. Modification can be as simple as sterilizing products required for the pharmaceutical industry to highly complex chemical modification to confer properties totally different from the native starch. A simple modification process is represented by washing, air classification, centrifugation and pre-gelatinization. Pre-gelatinization can be done in many forms from boiling in crude pots to drum dryers to modern multi-screw extruders.

Modified starch products are used in the food, paper and textiles industries. Modified starches increase the acceptability and palatability of many processed foods to consumers. Modified starches are also used to reduce costs of established food products. More expensive ingredients such as tomato solids, fruit solids, or cocoa powder can be extended with combinations of modified food starches, flavours, and other inexpensive food substances. Modified starches and dextrins have successfully replaced caseinates in meat emulsions, coffee whiteners, and imitation cheeses.

Modified dextrins are also used to replace butterfat in ice cream and ice milk, vegetable oil in salad dressings and shortening in icings. Modified food starches and dextrins play a very important role in cost reduction efforts. However, Africans have not started to carry out extensive research and applications on modified starches from cassava. Cassava starch finds good use in manufacture of noodles. Cassava based modified starch can be used as a fat mimetic in dairy systems due to its bland flavour. A low-fat product can be prepared with the organoleptic and textural properties of a traditional fat containing product. Cassava starch and cassava roots are being used in Malaysia and some other countries for the production of yeasts for animal feed, the human diet and for bakery yeast. Medium high glucose syrup - 63 DE - replaces sugar in marmalade and jam. Starch products are used as crystal and texture controller in Ice cream. *Cassava* speciality dextrin's replaces from 20% to 40% of gum Arabic in some hard gum candies.

Some Products from Modified Cassava Starches

Monosodium Glutamate (MSG).

This product is used extensively in many parts of the continent in powder or crystal form as a flavouring agent in foods such as meats, vegetables, sauces and gravies. Cassava starch and molasses are the major raw materials used in the manufacture of MSG in the Far East and Latin American countries. The starch is usually hydrolysed into glucose by boiling with hydrochloric or sulfuric acid solutions in closed converters under pressure. The glucose is filtered and converted into glutamic acid by bacterial fermentation. The resulting glutamic acid is refined, filtered and treated with caustic soda to produce monosodium glutamate, which is then centrifuged and dried in drum driers. The finished product is usually at least 99 percent pure.

Quality Control of Cassava Products

In the processing of cassava, questions naturally arise regarding efficiency and output; moreover, in selling the product the determination of quality becomes important. These problems can only be resolved by qualitative and quantitative study of the composition of the raw materials and the properties of the finished products. The financial return, especially in large factories, will depend to a certain degree on such control analyses, which in a way are actually part of the processing itself.

Analysis of Basic Materials

The two important basic materials requiring analysis are the cassava roots and the water used in processing. The best practical qualitative test of these materials consists in reproducing the whole process on a small scale and judging the resulting flour by comparing it with a standard sample or by analysing it according to the methods described farther on. In fact, for judging the suitability of the water available, small-scale processing is the only test of practical value. Apart from this, since starch is the substance to be isolated, a determination of the starch content in both the fresh roots and the pulp remaining after rasping and wet-screening is necessary for control of the efficiency of the process and in particular for determination of the rasping effect. Finally, tests for the presence of hydrocyanic acid are necessary owing to the important food uses of cassava.

Test Processing on a Small Scale

A random sample of, say, 10 kg of cassava is thoroughly washed to remove the cork layer; then either the whole or the peeled roots are grated or ground. The pulp is washed out over 50-mesh bronze gauze and the flour milk obtained over 260-mesh gauze. When the suspension reaches 3° Brix(approximately 35 g of dry starch per liter), it is left to settle for four hours. The top liquid is then decanted, and water is added to the settled starch to make slurry of 10° Brix, which is strained over 260-mesh gauze and left to settle for the second time. After decanting, the starch is mixed with water to a thick suspension (45° Brix) and filtered on a Buchner funnel under vacuum. The moist starch is dried in an oven, preferably in circulating air, commencing at a temperature of 50°C, and concluding at 60°C. The dried starch is sieved through silk before examination.

Determination of Starch Content in Fresh Roots and Waste Pulp

This analysis is best carried out with oven-dried pulp, a separate sample of fresh, moist pulp being used to determine moisture or water content, or with a sample of the fresh root material.

1. Quantitative determination of starch content is based on hydrolysis with acid and measurement of the resulting glucose. The weighed sample representing about 2.5 g of dried materials is ground and stirred with 250 ml of water for an hour, after which the insoluble residue is transferred to a vessel along with an additional 250 ml of water. After adding 200 ml of 0.5 NHCI, the solution is boiled under reflux and cooled; then the acidity is adjusted to pH 5 with NaOH, and when the solution reaches 250 ml, it is filtered. The glucose equivalent is determined by an aliquot according to the Munson and Walker method or any other suitable method (as described, for instance, in Methods of analysis of the Association of Official Agricultural Chemists, 1956). The amount of glucose multiplied by 0.93 is taken to equal the amount of starch which was in the aliquot.

2. A short-cut in the analysis of fresh roots is possible by determining the water content rather than the starch content. The method is practical especially where the variety of cassava and the growing conditions may be considered practically constant. The following empirical relation has been established as the result of a series of analyses of roots of four different strains of cassava, all grown on the same soil and during the same period.

   Whole root: Percent starch = percent total dry matter -7.3 = 92.7 - percent water

   Peeled root: Percent starch = percent total dry matter - 6.8 = 93.2 - percent water

3. A simple and inexpensive method for the quantitative determination of starch in cassava roots has been described by Krochmal and Kilbride. During peak seasons cassava tubers are kept in polyethylene bags and stored in deep-freeze. The frozen samples are sliced and blended with 500 ml of water for five minutes in a blender. The pulp is washed on a sieve with an additional 500 ml of water, and the fibrous material retained on the sieve is thrown away. The washed material is poured into aluminium pans and dried at about 85°C for 6-12 hours until a constant weight is attained. The weight of the residue represents the percentage of starch calculated from the weight of the sample.

Rasping Effect

The fraction of the starch in the roots which is set free by rasping may be evaluated directly by washing out a weighed sample of the pulp, as obtained from the factory rasper, on a 260-mesh sieve, collecting the starch on a filter, and weighing it after thorough drying. The percentage of free starch thus obtained divided by the total starch content of the pulp gives the rasping effect (R).

The following method for the determination of R may be preferable in certain respects as it obviates the rather difficult direct determination of the free starch by using only the analysis for starch and fiber content of the roots and the waste pulp produced in processing them.

If the starch contents of the roots and the waste pulp are s_r and s_w respectively and the corresponding fibre contents are f_r and f_w. it is readily seen that the fraction of the starch which remains bound to the fibre (i.e., occluded in cells which have escaped crushing by the rasper) amounts to $(s_w/f_w)/(s_r/f_r)$ and thus the rasping effect is $R = (1 - (s_f f_r / s_r f_w))$ X 100 percent

Prussic (Hydrocyanic) Acid Analysis

Qualitative Test (Quignard's Test)

Prepare sodium picrate paper by dipping strips into I percent picric acid solution and then, after drying, dipping them into 10 percent sodium carbonate solution, thereafter drying them again. Preserve these strips of paper in a stoppered bottle. Chop finely a small amount of the roots to be tested and put the choppings into a test tube. Insert a piece of moist sodium picrate paper, taking care that it does not come into contact with the root pulp. Add a few drops of chloroform and stopper the tube tightly. The sodium picrate paper gradually turns orange if the root material releases hydrocyanic acid. The test is a delicate one, and the rapidity of the colour change depends on the quantity of free hydrocyanic acid present.

Quantitative Determination (Alkaline Titration Method)

Put 10 to 20 g of the crushed root material in a distillation flask, add about 200 ml of water and allow to stand two to four hours, in order to set free all the bound hydrocyanic acid, meanwhile keeping the flask connected with an apparatus for distillation. Distil with steam and collect 150-200 ml of distillate in a solution of 0.5 g of sodium hydroxide in 20 ml of water. To 100 ml of distillate (it is preferable to dilute to a volume of 250 ml and titrate an aliquot of 100 ml) add 8 ml of 5 percent potassium iodide solution and titrate with 0.02 N silver nitrate (I ml of 0.02 N silver nitrate corresponds to 1.08 mg of hydrocyanic acid) using a microburet. The end point is indicated by a faint but permanent turbidity which may be easily recognized, especially against a black background.

Criteria for Quality of Flour and Starch

For a product like starch, which is used as a basic material in many quite different branches of industry, the value of a certain brand greatly depends on the purpose for which it is intended. Quality in cassava can therefore only be defined with reference to the end use. In each industry using the starch, the starting material will be subjected to certain tests in order to determine whether it is suitable for the process concerned. In some cases, a mere superficial test of purity will suffice (when, for instance, it is used as filler); in others, more elaborate determinations will be necessary. The value of the product in question will thus vary from case to case, and quality as well as price will emerge as a result of these investigations.

The analysis of cassava flour consists of a group of selected tests, which together provide the best possible general insight into the usefulness of the material. The analysis comprises chemical determinations, such as those of water, pulp and ash, as well as physiochemical tests for the measurement of viscosity and acidity. On the basis of the results of these tests, quality is usually designated in the form of a grade, that is, a cipher expressing the quality in general or, more specifically, in relation to a certain property. The letters A, B and C thus often denote first, medium and poor quality, each classification being bound to specified limits of the properties investigated.

The tests to determine the quality of cassava starch include those for mesh size, colour, odour, cleanliness, pulp or fibre content, moisture content, ash, acidity and viscosity of cold flour slurry as well as cooked starch paste. All these tests help establish the grade and therefore the commercial value of the product. The discussion of these quality features will bring out more clearly the significance of the properties involved. For several of these properties alternative methods of determination sometimes give valuable complementary information on the quality of the flour.

Mesh Size

This test measures efficiency of bolting. Fine pulp, however, obtained from bolting with disintegrators will pass the screens. In judging the purity of flour, neither this test nor the determination of pulp under the TIA system is sufficiently precise; it is supplemented in a way by the cleanliness test, and an additional determination of pulp content by hydrolysis would seem advisable.

Dry Appearance

In this test the brightness or whiteness of the flour is visually compared with that of a "standard," which is a prime-grade flour produced by certain first class factories. The result cannot be clearly expressed by enumeration. Besides, the difficulty of procuring flour of standard whiteness which will keep for a sufficiently long time is an important drawback.

With the many excellent modern types of spectrophotometric apparatus available, however, it would not be hard to devise an objective and accurate quantitative expression for the whiteness of the flour as measured by its reflectivity relative to that of a sufficiently durable standard of whiteness (e.g., barium sulfate). In fact, this method has been adopted in Indonesia. Remarkably enough, comparative experiments in which the same flours were judged by the direct visual method and by the objective reflectivity method have shown that the former test is about as accurate as the latter, provided the observer has enough experience with the visual method and has accustomed himself to the standard of whiteness adopted.

As an independent measure of the purity of the flour, dry appearance has lost much of its significance, but it is still a commercial criterion. It is customary in commerce to estimate the number of specks occurring on a flattened surface of the dry flour as an indication of clean processing.

Cleanliness

The following is a somewhat elaborate form of the test for specks which indicates the total amount of foreign particles in a sample. Five milliliters of distilled water are added to I g of dried starch. The mixture is stirred, and then 5 ml of 0.7 N sodium hydroxide solution are added and the uniform gelatinized mixture is examined for impurities. The degree of whiteness and clearness depends on the quantity of pigment, dirt and protein present in the starch.

Pulp

The amount of pulp, or fiber, present in the finished product is of foremost importance in deciding the usefulness of flour in various applications. The presence of insoluble cellulose is a serious hindrance in almost any industry where solutions of gelatinized starch are needed. Exceptions are the manufacture of corrugated cardboard and of plywood where the fiber is useful to a certain extent as a binder.

In the form given in the specification system under consideration, the test makes possible the determination of small amounts of fiber with comparative ease. The sediment volume measured is somewhat dependent on the fineness of the fiber. The presence of a slight trace of fiber, pulp or other impurity can be detected by microscopic examination of the size and shape of the starch granules.

The actual amount of cellulosic fiber in the flour and of foreign insoluble material can be determined by weighing the residue after a mild acid hydrolysis of the sample. Two to three grams of flour are boiled with 100 mm of 0.4 percent hydrochloric acid for one hour. The liquid is filtered through a weighed filter crucible fitted with filter paper or through a Jena glass filter, G 3. After washing with hot water the crucible is dried at 105°C to 110°C to a constant weight. One hundred times the gain in weight of the crucible divided by the weight of the test portion is the percentage of fiber and impurities.

It has been found that the determination of fiber content with these two methods runs parallel to a certain extent - that is, a 0.6 percent fiber content by the hydrolysis method corresponds to 10 ml of pulp per 50 g of flour by wet-screening.

A rough estimate of the amount of pulp may be based on the "crunch" of the flour - that is, the sound emitted when a sample, packed tightly in a small bag, is pinched between the fingers. "Crunch" is strong in pure flours, but above certain pulp content it is lost.

Viscosity

Raw starch suspended in water gives rise to more or less viscous slurries. While in some applications the viscosity of these suspensions may be of some technical importance, the term "viscosity of flour" is generally used for the viscosity of a solution of flour after gelatinization, because it is in that form that flour is used in most industries.

Numerous methods are used to determine this property. They differ in the instrument applied in the actual measurement of the rate of flow of starch solutions, and in the method of preparing the starch solutions to be tested. As the comparative test in the present specification system is rather subjective, a few quantitative determinations which have found wide application in the starch industry (particularly cassava)

Ash

The amount of inorganic constituents present, as measured by the ash content, can be considered an indication of clean processing and, in conjunction with the acid factor, conveys an impression of the quantity of metal ions bound to the raw starch. The colour of the ash is also of interest, as an off-colour indicates the presence of objectionable elements (e.g., brown-red from iron).

Moisture

Determination of moisture by the oven method, though simple to perform, requires many weighings and is otherwise rather lengthy. Moreover, the results tend to be low in a highly humid surrounding atmosphere such as is likely to occur in tropical regions.

A more rapid method, free from these objections, consists in boiling a sample with xylene (boiling point 135°C) and collecting the water driven out in the form of vapour, which separates from the xylene after condensation in a graduated tube. In the oven method, the starch sample is placed at a depth of less than 2 mm from the bottom in a receptacle with a tight-fitting cover and weighed. With the cover removed' it is heated in a vacuum oven at 105°C under a pressure of about 25psi.

Sweet potato processing and value addition

Table 26: Nutrient Value of Sweet Potatoes as Compared with other crops

Name of tuber crops	Nutrients per 1000g edible portion			
	Calories K.cal	Protein Gm	Vitamin C mg	Carotene I.U
Cassava	149	1.2	31.0	-
Coco yam	102	1.8	8	-
Yellow Guinea	71	1.5	10^2	20^2
Sweet Potato (Rare Variety)	121	1.6	37	75
Sweet Potato (Yellow Variety)	121	1.6	37	1,255
Sweet Potato (Deep Yellow) Orange-Flesh	121	1.6	37	2,400
Irish Potato	82	1.7	21	25

Source: Human Nutrition In Tropical Africa FAO, 1970

Processing

Both traditional and improved processing equipment were used in the processing of sweet potato flour. Locally manufactured and improved processing equipment were used in the processing of sweet potato flour in the study area. The processing equipments are including peeler, grater, dryer, and grinder.

Processing Method

There are two (2) basic processing techniques with seven processing stages involved in sweet potato processing. These include

a) The Traditional Processing Technique

b) The Improved Processing Technique.

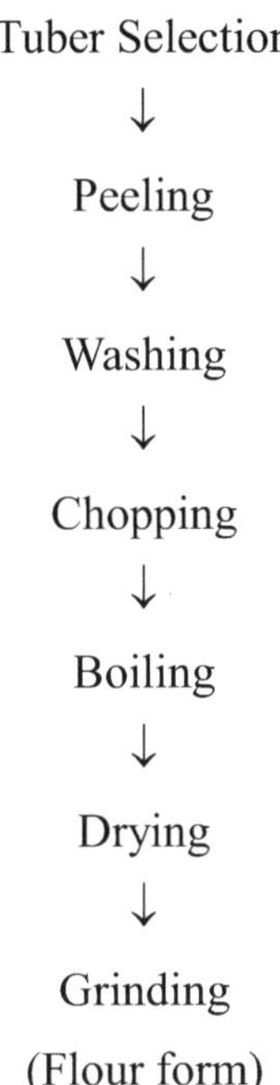

Figure 32: Flow diagram for traditional processing technique

The sweet potato processors usually process sweet potato flour with two methods. The first method is the **cold method** and the second method is the **hot method.**

1. Cold Processing Method

This involves peeling, washing and boiling of sweet potato tubers. The boiled tubers are put inside cold water and left to stand in cold water for 3 – 4 days. The water is then drained and washed to reduce the smell. The boiled tubers are dried and then ground.

2. Hot Processing Method

After peeling, washing and boiling, the hot tubers are soaked in hot water for 2 to 3 days. This helps to produce better and smoother flour. This flour can be prepared into paste and mixed with cassava flour and consumed with different soups.

3. Improved Processing Technique

The improved processing technique is still used in the study area. The improved sweet potato processing technique involves the use of Modern equipment. The stages include selection and weighing of tubers, washing, peeling, chopping / slicing, drying, primary grinding, attrition grinding and suction blowing / bagging

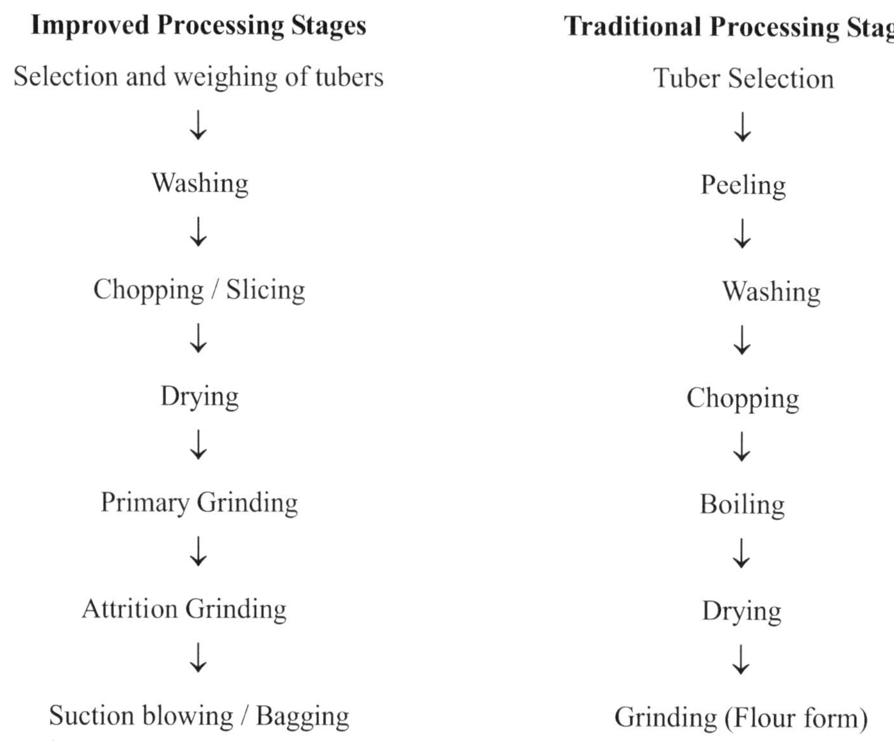

Figure 33: Process comparison between Improved and Traditional Processing Stages of Sweet Potato flour

The tubers are allowed to remain in the ground for a week after cutting the top of the vine to quicken the sweetening process. These tubers are graded by removing the infected ones. The selected tubers are then weighed on the platform scale. Peeling is done by Rotary Screen Friction Peeler, which washes and peels at the rate of 650 kg fresh tubers per hour. This is followed by the chopping / slicing. Tubers are then selected and fed to a chopper or slicer then followed by drying. This is done with the use of dryer, which is horizontal in shape. The Primary Grinding and or Attrition Grinder are used for grinding. This followed by blowing and the bagging.

Sweet Potato recipes

1. Sweetpotato Porridge

Ingredients

Sweetpotato flour 1 heaped table spoon

Millet flour 4 heaped table spoon

Soya flour 1 heaped table spoon

Lemon 1 small

Sugar 2 table spoons

Water 6 cups

Figure 34: Value added products from Sweet potato

Procedure

1. Bring five cups of water to boil.
2. Mix the flours and make a paste with the remaining one cup of water.
3. Pour the paste into the boiling water and keep stirring to prevent lumps.
4. Make juice from the lemon while pot continues to boil for 20 minutes.
5. The cooked product should jell
6. Remove from fire add lemon juice and sugar.
7. Cool, then serve warm.

2. Sweetpotato Crisps

Ingredients

Sweetpotato roots 6 medium

Oil 2 cups

Salt to taste

Water 2 containers

Procedure

1. Remove soil from roots and peel as you place in clean water.
2. Wash off any soil.
3. Slice into very thin pieces using a knife or larger blade of grater.
4. Drain off the water.
5. Heat the oil and deep fry till starting to brown.
6. When brown remove and drain.
7. Salt and serve warm or cold.

3. Sweetpotato "Mshenye"

Ingredients

Sweetpotato roots 10meduim

Maize 2 cups

Beans 4 cups

Salt to taste

Water adequate

Procedure

1. Sort maize and beans and pre-soak for 6-8 hours.
2. Boil the maize and beans till almost cooked.
3. Remove soil from sweetpotato roots and peel.
4. Wash and slice the sweetpotato roots into desired shapes.
5. Add the sliced sweetpotato roots to the maize and beans and let cook.
6. When sweetpotato roots are soft and maize and beans well cooked mash.
7. Add salt to taste and serve as balls heaped on plates.

4. Sweetpotato Relish

Ingredients

Sweetpotato leaves 1 kg

Onions 2 medium

Tomatoes 4 medium

Flavour 4 tablespoons

Oil/Fat 4 tablespoons

Salt 1 tablespoon

Warm water ½ container

Procedure

1. Clean leaves by removing dirty and very old ones.
2. Prepare the onions and tomatoes and slice into separate dishes.
3. Shred the leaves.
4. Wash twice in warm water to remove the anti-nutrients.
5. Heat the oil and fry onions till they start to brown.
6. Add tomatoes and let cook for a while.
7. Add the vegetables and let cook for 5 minutes
8. Add the flavour and stir the contents and let cook till done.
9. Serve with bananas/ugali or kaunga, sima or nshima/rice.

   **The flavour can be alternated with milk, groundnut paste, coconut milk, soya flour etc.*

5. Sweetpotato Soya Chapatti

Ingredients

Grated sweetpotato 1 cup

Wheat flour 2 cups

Soya flour 1 cup

Salt 1 teaspoon

Lukewarm water adequate

Oil ½ cup

Procedure

1. Mix dry ingredients together in a bowl.
2. Add the grated sweetpotato and mix.
3. Add 1 tablespoon of oil to the flour and mix well.
4. Add the water to the mixture in the bowl and knead till stiff smooth paste is formed.
5. Divide the dough into 8-10 equal balls.
6. On a floured surface roll one ball at a time.
7. Fold each ball at a time to form a strip.
8. Coil each strip to form a circle and put aside for 20 minutes
9. On a floured surface, roll out each coil into a thin circular sheet.
10. Grease a shallow frying pan.
11. Fry each circular sheet on both sides till golden brown.
12. Ensure to grease both sides.
13. The product is the chapatti and can be served with stew or sauce or tea.

6. Sweetpotato Mandazi

Ingredients

Sweetpotato mash ½ cup

Wheat flour 2 cups

Sugar 2 table spoons

Salt pinch

oil 2 cups

Baking powder 1 table spoon

Lukewarm water adequate

Procedure

1. Put the sweetpotato mash in a mixing bowl and sift in the dry ingredients.
2. Add water and mix into dough.
3. Knead the dough well while adding 2 tablespoons of oil.
4. On a floured surface, roll the dough to about 1 cm thickness.
5. Cut into desired shapes.
6. Deep fry while turning till golden brown.
7. Remove from oil, drain and serve warm or cold.

7. Sweetpotato Doughnuts

Ingredients

Grated sweetpotato ½ cup

Wheat flour 2 cups

Yeast 1 teaspoon

Sugar 2 tablespoon

Salt pinch

Oil 2 cups

Milk/Egg optional

Cooking fat 1 tablespoon

Lukewarm water adequate

Procedure

1. Put yeast and 1 tablespoon sugar in a cup
2. Add 3 tablespoon of warm water and leave for 10 minutes to rise.
3. Put the grated sweetpotato into a mixing bowl and sift in the dry ingredients.
4. Rub in the cooking fat and then add the risen yeast and mix.
5. Add water to the mixture and knead into a dough
6. On a floured surface, roll the dough slightly.
7. Make dough into a ball and return to mixing bowl.

8. Cover the mixing bowl with wet warm cloth and leave to double.
9. Re-knead the dough after doubling and roll onto a floured surface.
10. Cut into desired shapes and deep fry till golden brown.

8. Sweetpotato Crackies

Ingredients

Sweetpotato flour 1 cup

Wheat flour 2 cups

Blue band 2 table spoons

Eggs 3

Salt 1 table spoon

Spices 1 table spoon

Oil 2 cups

Baking powder 2 tea spoons

Procedure

1. Sift all dry ingredients in a mixing bowl.
2. Add blue band and rub in.
3. Whisk the eggs and add to the contents in the bowl.
4. Knead to a smooth dough, if hard add a little warm water.
5. Pack dough in noodle machine.
6. Heat oil and drop contents in machine by turning the handle round.
7. Cut contents from machine with a knife to reduce the size.
8. Let cook till brown
9. Remove and drain oil and let cool in a covered container.

9. Sweetpotato Onion Bites

Ingredients

Sweetpotato mash 1 cup

Wheat flour 2 cups

Baking powder 3 tsp

Chilli pepper 3/4 tsp

Salt 1 tsp

Spring onion leaves 1/2cup

Water

Cooking fat 1 tsp

Procedure

1. Sift all dry ingredients in a mixing bowl. Pound the onions
2. Add the sweetpotato mash and cooking fat and mix well to a dough
3. Add a little water at a time and knead to a light texture, let it relax for 10 - 15 minutes
4. Heat oil in a pan
5. Make small sized balls and drop them into the hot oil
6. Cook till brown drain and serve.

10. Sweetpotato Strips

Ingredients

Sweetpotato flour 1 cup

Soya flour 1/2 cup

Wheat flour 2 cups

Cooking fat 4 table spoons

Eggs 1

Sugar 3 table spoons

Oil 2 cups

Baking powder 3 tea spoons

Procedure

1. Sift all dry ingredients in a mixing bowl
2. Add cooking fat and rub in.
3. Beat eggs and add to the bowl and mix.
4. Add a little warm water and knead to smooth dough.
5. Roll dough on floured surface to 1centimeter.
6. Cut small strings from rolled dough.
7. Heat oil and drop in the strings and let cook.

8. When brown remove and drain.
9. Keep in covered container to prevent hardening.

11. Sweetpotato Bread

Ingredients

Grated sweetpotato ½ cup

Wheat flour 2 cups

Yeast 1 teaspoon

Sugar 1 tablespoon

Salt pinch

Luke warm water adequate

Oil 2 tablespoons

Procedure

1. Mix yeast and sugar in a cup.
2. Add 3 table spoons of water to the cup and leave to rise.
3. For quick rising cover cup with a warm cloth and put in sun for 5 minutes or 10 minutes at room temperature.
4. Mix grated sweetpotato with other dry ingredients in a mixing bowl.
5. Add the yeast mix and water into mixing bowl.
6. Knead into dough and add the oil to make it smooth.
7. Divide into two parts.
8. Grease bread tins and shape each dough and place in tin.
9. Leave to rise to double size.
10. Bake in oven at 200°C (400°C) for 15 - 20 minutes.
11. Remove and allow cooling and then wrapping.

12. Sweetpotato Buns

Ingredients

Sweetpotato mash 1 cup

Wheat flour 3 cups

Sugar 2 tablespoons

Salt pinch

Yeast 1½ teaspoons

Oil/Fat 3 tablespoons

Water adequate

Procedure

1. Put yeast with 1 teaspoon sugar in a cup,
2. Add 2½ tablespoons of warm water and leave to rise.
3. Put the mashed sweetpotato in a mixing bowl and sift in the dry ingredients
4. Add oil/fat and rub in till it crumbles.
5. Add risen yeast and mix.
6. Add water and knead till done to required texture.
7. Roll into a ball, put into mixing bowl and cover with wet cloth or put into an oiled poly bag and let double.
8. Knead the doubled dough.
9. Divide dough into equal small balls and roll out to make desired shapes.
10. Put in oiled baking pan and leave at room temperature for 10 minutes.
11. Bake for 20 minutes at 170°C or 350°F or till crust is golden brown.

13. Sweetpotato Cake

Ingredients

Sweetpotato mash 1 cup

Wheat flour 3 cups

Eggs 4

Blue band 5 tablespoons

Baking powder 3 teaspoons

Lemon 1 medium

Sugar 3 table spoons

Procedure

1. Sift all dry ingredients in a bowl.
2. Add the sweetpotato mash and 4 tablespoons of blue band and rub in.
3. Beat the eggs and add to the bowl and mix well.

4. Grate lemon rind and add to the bowl and mix.
5. Make juice from the lemon and add to the bowl contents and mix well.
6. If consistency not runny, add a little water.
7. Grease baking pan and pour in contents.
8. Bake in oven at 175°C (360°f) for 30 minutes or till brown.
9. Alternatively bake on open fire (see tips on open-fire baking below)

Tips on Open-Fire Baking

1. Pre heat the charcoal stove /jiko, Kenya Ceramic Jiko (KCJ) or maendeleo stove.
2. Grease a heavy pan with lid.
3. Pour mix dough contents into the pan.
4. Cover the pan with lid preferably heavy chapati pan.
5. Remove fire from stove and place on the lid evenly.
6. Leave very little fire in the fire – box and cover with ash.
7. Place covered pan with fire on the ash covered stove.
8. Keep fire on lid burning by adding twigs for 2 minutes
9. Let cook for another 30 – 40 depending on type of charcoal used minutes.
10. Remove lid with fire, test cake with knife by piercing in the middle.
11. If done knife should be dry - if not done knife will be wet with uncooked contents.
12. If done remove and cool cake on rack.
13. If not done return and replace lid with fire for a while then remove.

14. Sweetpotato Biscuits

Ingredients

Sweetpotato mash 1 cup

Wheat flour 2 cups

Sugar 3 table spoons

Salt pinch

Blue band 3 table spoons

Baking powder 1 tsp

water 2 cups

Procedure

1. Sift the dry ingredients in a mixing bowl.
2. Add the blue band and rub in till the mixture crumbles.
3. Add water, knead to a stiff paste
4. Roll out on a floured board.
5. Cut into shapes and arrange on a greased baking pan.
6. Prick with a fork to prevent dough from rising.
7. Bake for 15 minutes at 175°C (350°F) or until evenly brown.

15. Sweetpotato Pineapple Upside Down

Ingredients

Sweetpotato flour 1

Wheat flour 3

Eggs 4

Blue Band 5 tablespoons

Baking powder 3 teaspoons

Sugar 3 table spoons

Pineapple 1 small ripe

Procedure

1. Sift all dry ingredients in a mixing bowl.
2. Cream the blue band and sugar in a small bowl.
3. Break eggs one at a time and pour into cream mixture.
4. Mix well then pour in the dry ingredients and fold in, to give a runny mix.
5. If hard add water or milk to make mixture runny and smooth.
6. Clean the pineapple and peel. Slice pineapple into 1 inch thick.
7. Grease baking tin and arrange the pineapple slices in.
8. Pour cake mix into tin and bake in oven at 176°c (360°F) for 30 minutes or till brown.
9. Remove and turn contents into a plate or small tray.
10. Serve as dessert/ sweet.

16. Sweetpotato Pie

Ingredients

Sweetpotato flour 1 cup

Wheat flour 2 cups

Fat 8 table spoons

Salt to taste

Rice 2 cups

Dhania 1 bunch

Minced beef ½ kg

Egg plants 1 large

Onions 1 medium

Tomatoes 2 medium

Mixed spices 1 teaspoon

Milk 1 cup

Egg 1

Procedure

1. Prepare dhania, onions and tomatoes and cut into separate containers.
2. Fry onions with 2 table spoons of oil.
3. Add tomatoes, let cook then add dhania.
4. Add minced beef and let cook for 20 minutes
5. When about ready add mixed spices and let cook for another 10 minutes.
6. Sort rice, wash once and bring 4 cups of water to boil.
7. Pour rice into boiling water, add a little salt then cook for 15 mins. or till there is no water.
8. Clean egg plants and cut into slices.
9. Use 2 tablespoons of fat to shallow fry the slices, and arrange in a greased baking tin.
10. Pour the rice on the arranged egg plants.
11. Pour the beef on the rice and evenly distribute.
12. Melt the remaining fat, reduce heat.
13. Fold the mixed flour into the melted fat, a little at a time.

14. Add milk to flour - fat mix and continue till all flour is used.
15. The mixture should spread when poured.
16. Pour the mixture onto the beef and spread evenly.
17. Beat the egg and spread on the pastry.
18. Bake in oven at 170°C (350°F) for 30 minutes.
19. Remove and serve as complete meal.

17. Sweetpotato Juice

Ingredients

Sugar 4 cups

Boiled peeled sweetpotato roots 8 medium sized roots

Citric acid/ lemon juice 3 teaspoon/ 5 fruits

Water boiled and cooled 5 liters

Fruit flavouring (optional) 1 drop or add tamarind to taste

Procedure

1. Boil water and sugar and then leave to cool.
2. Mash boiled sweetpotato or blend, mix the product with the boiled water and then sieve.
3. Add citric acid/ lemon juice and fruit flavour if desired and mix well.
4. Pour into a jug, chill if possible and serve cold as fresh juice.

18. Sweetpotato Flours

The number of different sweetpotato flours for use in various products, Sweetpotato flour with optional additives of tamarind juice, lemon juice or dried powder made from sun dried slices of unripe mangoes.

Sweetpotato Composite Flour for Making Bread

Mix sweetpotato flour with cassava and sorghum flour at weight ratios of 2:2:1. As above it can be enhanced with optional additives.

Sweetpotato Composite Flour for Making Porridge

Mix sweetpotato flour with maize flour at a weight ratio of 2:1. As above it can be enhanced with optional additives.

References

Addy, N.D. and Stuart, D.A. 1986. Impact of biotechnology on vegetable processing. Food Technol. 40(10), 64-66

Arthey, D. and Dennis, C. 1991. Vegetable Processing. Chapman & Hall, London, New York

Balagopalan,C., Ray,R.C., Moorthy, S.N., Lila Babu, Shanthi,B. and Sheriff, J.T. 2001. Integrated technologies for value addition and post harvest management in sweet potato *Ipomea batatas* (L.) Lam: The indian experience. Journal of Root Crops.27(2):497-507.

Choudhury, B. Vegetables, National Book Trust, India, 1977

Jyothi, A. N., Sheriff, J. T. and Sajeev, M. S. 2009. Physical and functional properties of arrowroot starch extrudates, *Journal of Food Science*, 74(2), E97-E104.

Jisha, S., Sheriff, J.T and Padmaja, G. 2010. Nutritional and functional properties of extrudates from blends of cassava flour with cereal and legume flours. International Journal of Food Properties. 13: 1002-1013.

Katyal, S.L. Vegetable Growing in India, Oxford and IBH Publishing Co., New Delhi, 1977

Luh, B.S. and Woodroff, J.G. 1988. Commercial Vegetable Processing, 2nd ed. Chapman & Hall, London, New York.

Nath, P. Vegetables for the Tropical Region, Indian Council and IBH Publishing Co., New Delhi, 1977

Shakuntala Many and M. Shadaksharaswamy: Foods: Facts and Principles, 2nd edn., New Age International (P) Ltd., New Delhi, 1987

Sheriff, J.T. and Balagopalan, C. 1999. Evaluation of a multi purpose starch extraction plant. Tropical Science 39, 147-152.

Shanavas, S., Padmaja, G., Moorthy, S.N., Sajeev, M.S and Sheriff, J.T. 2011. Process optimization for bioethanol production from cassava starch using novel eco-friendly enzymes. Biomass and Bioenergy. 35: 901-909.

❑❑❑

Chapter - 5

Minor Forest Products Based Foods

Tamarind *(Tamarindus Indica)*

Introduction

The fruit of tamarind a pod 5 to 15 cm long, 3 to 10 seeds surrounded with edible pulp which is principal souring agent for sauces, chutney, in beverages and in general cooking. Pulp is carminative, laxative, given as infusion in biliousness and febrile conditions. It is also used in dyring and tanning and for polishing and cleaning metal ware. The tartaric acid is extracted from unripe fruits. The polysaccharide (jellose) is extracted from seeds, which is used as a sizing material in the cotton and jute industries. Besides, polyose obtained from the seed is good substitute for fruit pectin in the preparation of jam, jelly or marmalade. The bark and leaves are used for tanning. Tamarind balls are prepared after taking out seeds from the fruit. Tamarind paste and tamarind juice concentrate are the other commercial products.

The fruits, flattish, bean-like, irregulary curved and bulged pods, are born in great abundance along the new branches and usually vary from 2 to 7 in long and from 3/4 to 1 1/4 in (2-3.2 cm) in diameter. Exceptionally large tamarinds have been found on individual trees. The pods may be cinnamon-brown or grayish-brown externally and, at first, are tender-skinned with green, highly acid flesh and soft, whitish, under-developed seeds. As they mature, the pods fill out somewhat and the juicy, acidulous pulp turns brown or reddish-brown.

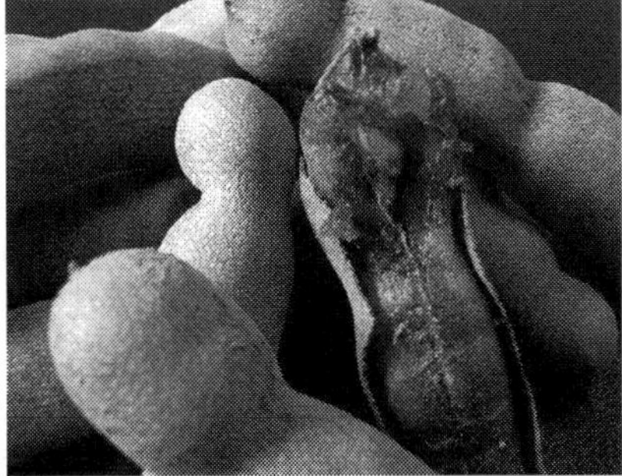

Figure 35: Tamarind pulp

Tamarind processing activity is very important in tribal areas and is done manually. There are plenty of trees available. Tamarind is a vital ingredient in many food items. Presently the tribs are not adopted any machinery or technology to process the tamarind. Tamarind paste can be prepared mechanically. It is soaked in a hot water and the juice is extracted manually from the resultant soft pulp, which is then used for cooking. Tamarind pulp contains 20% moisture, 6% Fiber and 67 % carbohydrates. Tamarind juice contains the mellow flavoured potassium acid tartrate, besides free tartaric acid .The concentrate is made free of fiber, seed etc.

Market Potential

There is wide awareness about the product and its properties as a ready to use food. Currently tamarind products are available in the market. The product can find a place in all towns and cities in super Markets and departmental stores. Export potential exists in countries like UAE, USA & Malaysia. Spices board and some other department have show initiative to promote this product in India.

Why process Tamarind Fruits

- The fruit is high in protein, carbohydrates, potassium, phosphorous and calcium and is a source of iron, vitamin C, thiamine and niacin.
- Processing increases the shelf life of the fruit
- Processing adds value and increase income

Pre – Processing

- Sundry fresh fruits or use small scale dehydrators
- Crack and separate pulp and fibers from the broken shells.
- Removed the seeds
- Store the pods for several weeks at 20°C or store the pulp for 4-6 months below 10°C by packing in high density polythene or store one year when the pulp is mixed with salt.

Processing and Value Addition

Tamarind Juice

Procedure

1. Boil tamarind pulp in water,
2. Filter juice to remove the pulp,
3. Pour into bottles and seal,
4. Cool rapidly to room temperature in cold water,
5. Pack in well- sealed clean glass or plastic, bottles
6. Store in a dark, cool place

Tamarind Concentrate

This can be easily dispersible in water, and can be used for many purposes, such as in ketchups, sauces, soft drinks, dairy products and as a souring agent.

Procedure

1. Soak tamarind pulp in water and boil.
2. Separate fine and pulpy matter using a filter
3. Press the residue and mix this with the extract,
4. Concentrate the filtered extract by evaporation under vaccum,
5. Fill containers, cool and seal
6. Store in airtight plastic or glass bottles or cans, in the dark, for over a year Coriander, cumin, cardamom, chillies, clove, cinnamon, mustard seed, turmeric powder

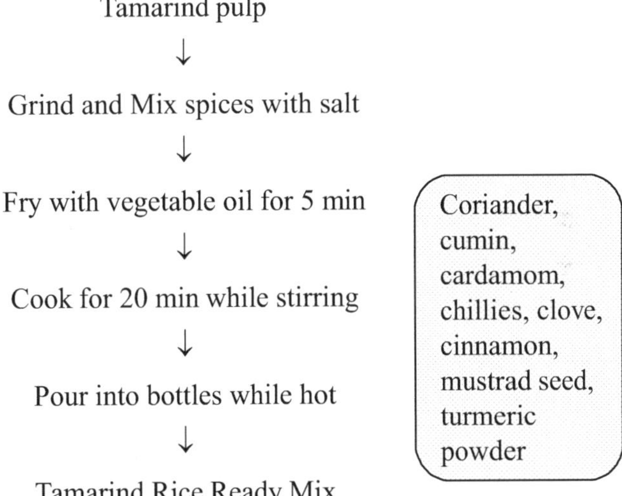

Figure 36: Flow diagram for Tamarind rice ready mix

Table 27: Potential of processed products from tamarind

Pickles/ Chutney	Jam / Jelly	Juice/ Nectar
Past/Concentrate	Candy/Leather	Fruit-Puree /Pulp

Tamarind Leather

Procedure

1. Prepare either citric acid or lemon juice dip; or sulphite dip by dissolving 6 grams metabisulphite in 10 liters of water. Dip tamarind pulp
2. Puree pulp and heat to 90°C (to inactive enzymes)
3. Adjust the sweetness and acidity by adding sugar and citric acid or lemon juice
4. Add nuts, spices and other flavorings (optional)
5. Spread in a thin layer on greased paper and dry

Pack rolls of leather, interleaved with greaseproof paper in moisture-proof, heat sealed bags, store in a cool, dark place for up to 9 months.

Table 28: List of machinery and equipment for tamarind processing unit

S.No	Description	Qty	Amount (Rs)
1	Pulper	5@ Rs 8000	40000
2	Kettle SS with Blender	1 @ Rs 24000	24000
3	Utensils	1	10000
4	Package machine	1 @ Rs 3000	3000
5	Electrification	1	3000
	Total		80000

Keeping Quality

To preserve tamarinds for future use, they may be merely shelled, layered with sugar in boxes or pressed into tight balls and covered with cloth and kept in a cool, dry place. For shipment to processors, tamarinds may be shelled, layered with sugar in barrels and covered with boiling syrup. East Indians shell the fruits and sprinkle them lightly with salt as a preservative. In Java, the salted pulp is rolled into balls, steamed and sun-dried, then exposed to dew for a week before being packed in stone jars. In India, the pulp, with or without seeds and fibers may be mixed with salt (10%), pounded into blocks, wrapped in palm-leaf matting, and packed in burlap sacks for marketing. To store for long periods, the blocks of pulp may be first steamed or sun dried for several days.

Medicinal Uses of Tamarind

Medicinal uses of the tamarind are uncountable. The pulp has been official in the British and American and most other pharmacopoeias and some 200,000 lbs (90,000 kg) of the shelled fruits have been annually imported into the United States for the drug trade, primarily from the Lesser Antilles and Mexico. The European supply has come largely from Calcutta, Egypt and the Greater Antilles. Tamarind preparations are universally recognized as refrigerants in fevers and as laxatives and carminatives. Alone, or in combination with lime juice, honey, milk, dates, spices or camphor, the pulp is considered effective as a digestive, even for elephants, and as a remedy for biliousness and bile disorders, and as an anti-scorbutic.

In native practice, the pulp is applied on inflammations, is used in a gargle for sore throat and, mixed with salt, as a liniment for rheumatism. It is, further, administered to alleviate sunstroke, Datura poisoning, and alcoholic intoxication. In Southeast Asia, the fruit is prescribed to counteract the ill effects of overdoses of false chaulmoogra, Hydnocarpus anthelmintica Pierre, given in leprosy. The pulp is said to aid the restoration of sensation in cases of paralysis. In Colombia, an ointment made of tamarind pulp, butter, and other ingredients is used to rid domestic animals of vermin

Tamarind leaves and flowers, dried or boiled, are used as poultices for swollen joints, sprains and boils. Lotions and extracts made from them are used in treating conjunctivitis, as antiseptics, as vermifuges, treatments for dysentery, jaundice, erysipelas and hemorrhoids and various other ailments. The fruit shells are burned and reduced to an alkaline ash which enters into medicinal formulas. The bark of the tree is regarded as an effective astringent, tonic and febrifuge. Fried with salt and pulverized to an ash, it is given as a remedy for indigestion and colic

A decoction is used in cases of gingivitis and asthma and eye inflammations; and lotions and poultices made from the bark are applied on open sores and caterpillar rashes. The powdered seeds are made into a paste for drawing boils and, with or without cumin seeds and palm sugar, are prescribed for chronic diarrhea and dysentery. The seed coat, too, is astringent, and it, also, is specified for the latter disorders. An infusion of the roots is believed to have curative value in chest complaints and is an ingredient in prescriptions for leprosy.

The leaves and roots contain the glycosides: vitexin, isovitexin, orientin and isoorientin. The bark yields the alkaloid, hordenine.

Bael Fruits

Importance

Vegetables and fruits provide health and nutrition promoting compounds in human diet. Their constituents prevent diseases through several mechanisms and thus increase one's life span and the quality of health and life. Among the underutilized fruits, Bael fruit (*Aegle marmelos*) occupies an important place. Its scientific name is Aegle

Marmelos, while it belongs to family Rutaceae and is commonly known as Bael, Bengal Quince, Indian Bael, Wood Apple, Matoom, etc.

Normally, the fruit is harvested when yellowish-green and kept for 8 days while it loses its green tint. Then the stem readily separates from the fruit. The fruits can be harvested in January (2 to 3 months before full maturity) and ripened artificially in 18 to 24 days by treatment with 1,000 to 1,500 ppm ethrel (2-chloroethane phosphonic acid) and storage at 86° F (30° C). Care is needed in harvesting and handling to avoid causing cracks in the rind.

Keeping Quality

Normally-harvested bael fruits can be held for 2 weeks at 86° F (30° C), 4 months at 48.2° F (9° C). Thereafter, mold is likely to develop at the stem-end and any crack in the rind.

Why Process Bael Fruits

- The fruit is rich in protein, carbohydrates and minerals and is a source of carotene, thiamine, riboflavin, niacin and vitamin C
- Processing reduce post harvest losses
- Processing increase the shelf life of the fruit
- Processing adds value and increase income

Nutrition Composition

Table 29: Nutrition Value Per 100 g of Edible Portion

Water	54.96-61.5 g
Protein	1.8-2.62 g
Fat	0.2-0.39 g
Carbohydrates	28.11-31.8 g
Ash	1.04-1.7 g
Carotene	55 mg
Thiamine	0.13 mg
Riboflavin	1.19 mg
Niacin	1.1 mg
Ascorbic Acid	8-60 mg
Tartaric Acid	2.11 mg

Figure 37: Whole bael fruit

Figure 38: Broken bael fruit

Table 30: Nutritive value comparison with amla fruit

	Energy (Kcals)	Moisture (g)	Protein (g)	Fat (g)	Mineral (g)	Fiber (g)	Carbohydrates (g)	Calcium (mg)	Phosphorus (mg)	Iron (mg)
Amla	58	82	0	0	0	3	14	50	20	1
Bael fruit	137	61	2	0	2	3	32	85	50	1

Bael has numerous seeds which are densely covered with fibrous hair and are embedded in a thick aromatic pulp. The flesh is eaten fresh or dried. The bael fruit consists of moisture 61.5 percent, protein 1.8 percent, fat 0.3 percent, minerals 1.7 percent, fiber 2.9 percent and carbohydrates 31.8 percent per 100 grams of edible portion. Its mineral and vitamin contents include calcium, phosphorus, iron, carotene, riboflavin, thiamin, niacin and vitamin C.

Food Uses

Ripe bael fruit is regarded as the best natural laxative. For best results, it should be taken in the form of sherbet, which is prepared from the pulp of the ripe fruit. After breaking the shell, the seeds are removed, with the contents spooned out and sieved. Milk and sugar are added to make it more palatable. The pulp of the ripe fruit can also be taken without the addition of milk or sugar. The unripe or half-ripe fruit is very effective in treating chronic diarrhoea and dysentery where there is no fever. Best results are obtained by the use of dried bael fruit or its powder. The bael fruit, when it is still green, is sliced and dried in the sun. The dried bael slices are powdered and preserved in airtight bottles. The unripe bael can also be baked and used with jaggery or brown sugar.

The pulp is often processed as nectar or "squash" (diluted nectar). A popular drink (called "sherbet" in India) is made by beating the seeded pulp together with milk and sugar. A beverage is also made by combining bael fruit pulp with that of tamarind. These drinks are consumed perhaps less as food or refreshment than for their medicinal effects.

Mature but still unripe fruits are made into jam, with the addition of citric acid. The pulp is also converted into marmalade or syrup, likewise for both food and therapeutic use, the marmalade being eaten at breakfast by those convalescing from diarrhoea and dysentery. A firm jelly is made from the pulp alone, or, better still, combined with guava to modify the astringent flavor. The pulp is also pickled. Bael pulp is steeped in water, strained, preserved with 350 ppm SO_2, blended with 30% sugar, then dehydrated for 15 hrs at 120° F (48.89^o C) and pulverized. The powder is enriched with 66 mg per 100 g ascorbic acid and can be stored for 3 months for use in making cold drinks ("squashes").

A confection, bael fruit toffee, is prepared by combining the pulp with sugar, glucose, skim milk powder and hydrogenated fat. Indian food technologists view the prospects for expanded bael fruit processing as highly promising.

Processing & Value Addition

Since bael is a seasonal fruit, processing of fruit and converting it into value added products has to be done during availability and predominantly market for the value added products generated thereof. Keeping in mind the benefits and application of post harvest technologies to the underutilized fruit like Bael a wide range of value added products are developed.

At the first stage, selection of fruit, nutritional analysis, economical method of extracting pulp followed by preserving the fruit pulp, shelf life study of bael fruit pulp and leaves are conducted.

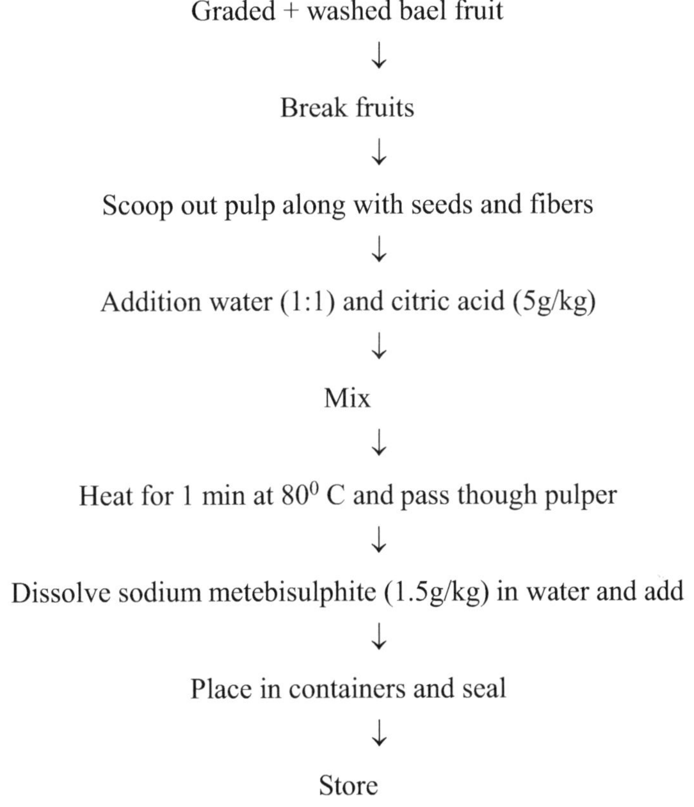

Figure 39: Flow diagram for pre-processing of bael fruit

Storage of Bael Fruits

- **Fresh fruits:** for 15days at 30° C, when harvested at full maturity (light green)
- For only 1 week at 30° C, when harvested tipe
- For 3 months at 9° C
- **Pulp:** for up to 6 months, when packed in heat- sealed containers

The bael fruit pulp can be stored up to four months in pet jars and bottles at room temperature. The bael leaf powder can be stored in both Low Density Polyethylene (LDPE) and Metalized Polyester Polypropylene (MPP) packages for six months.

Bael Fruit Pulp Incorporated Products

Cereal Bar

- Cereal bar is developed using puffed cereals and millets in two combinations as (corn and sorghum) with the incorporation of bael fruit pulp at 25%, 30%and 35%.
- In both the combination 35% incorporation was highly acceptable.

Nutritional Composition

Corn incorporated cereal bar contains

- 13.1 g protein,
- 964.50 mg _-carotene,
- 5.34 mg vitamin-C
- 2.5g crude fiber and

Sorghum incorporated contains 17.3g protein, 964.50 and 3.0g crude fiber.

Bael Toffee

- The toffee is developed with the incorporation of bael fruit pulp at 50%, 75% and 100%
- Among the different incorporation tried and tested 50% was highly acceptable

Nutrition Composition

The 50% incorporated product contains 74.g protein and 2.1mg of vitamin-C.

Bael Squash

Bael squash was developed with the incorporation of

1. Papaya fruit pulp
2. Sweet lime at different levels
3. Bael squash was developed without incorporation of other fruits.

The organoleptic evaluation showed that 20% incorporation was acceptable.

Nutrition Composition

Sweet lime incorporated squash has 1% acidity and 1.33mg vitamin incorporated squash has 2.1mg vitamin acidity and 2.5 mg vitamin-C.

Uses and Benefits of Developed Products

- Bael leaf incorporated roti mix has a lot of medicinal values for diabetic, heart diseases, suffocation and asthma, etc., This roti mix can be used by all the age groups including old age and obese person.
- Bael squash is good for all the groups and it helps to keep the body cool and quenches thirst.
- Bael cereal bar and toffee with high nutritive value can be given to the children as a substitute for traditional toffee.

Stevia

In spite of the prominence, Stevia has obtained as a calorie free sweetener and flavour enhancer, it contains a variety of constituents besides the steviosides and rebaudiosides. This including the nutrients specified above and a good deal of sterols, triterpenes, flavonoids, tannins, and an extremely rich volatile oil comprises rich proportions of aromatics, aldehyde, monoterpenes and sesquiterpenes.

Stevia has medicinal properties, too. Scientific research has shown it to be beneficial in regulating blood sugar levels, bringing them into normal range. It is also used as a digestive aid. As a skin care product, it has been used to clear blemishes, tighten skin to remove wrinkles, to heal mouth sores and to treat a variety of wounds. It has also been used to treat eczema, seborrhea and dermatitis.

Hypoglycaemic Action

Stevia is helpful for hypoglycemia and diabetes because it nourishes the pancreas and thereby helps to restore normal pancreatic function.

Cardiovascular Action

The long-term use of Stevia would probably has a cardiotonic action, that is, would produce a mild strengthening of the heart and vascular system.

Antimicrobial Action

The ability of Stevia to inhibit the growth and reproduction of bacteria and other infectious organisms is important. Stevia has even been shown to lower the incidence of dental caries.

Digestive Tonic Action

Stevia made a significant contribution to improved digestion, and that it improved overall gastrointestinal function. Stevia tea, made from either hot or cold water, is used as a low calorie, sweet tasting tea, as an appetite stimulant, as a digestive aid and as an aid to weight management.

Honey

Introduction

Honey consists of a mixture of sugars, mostly glucose and fructose. In addition to water (usually 17-20%) it also contains very small amounts of other substances, including minerals, vitamins, proteins and amino acids. A minor, but important component of most types of honey is pollen. These components contribute to the different flavours that honey can have, and make honey a nutritious food that has a high demand in many regions of the world.

Bees make honey mainly from the nectar of flowers, but they also use other plant saps and honeydew. As a bee sucks the liquid up through its proboscis and into its honey sac, it adds a small amount of enzymes, and some of the water in the nectar is evaporated. The enzymes convert sugars in the nectar into different types of sugars; honeys always contain a wide range of sugars that vary according to the nectar source. The bees then place the liquid nectar into cells in the honeycomb. The temperature inside the hive is usually around 35°C and, together with ventilation caused by bees fanning their wings, this temperature causes further evaporation of water from the nectar. When the water content is less than 20%, the bees seal the cell with a wax capping. The honey is now 'ripe' and will not ferment.

Many species of bees collect nectar which they convert into honey and store as a food source. However, only bees that live together in large colonies store appreciable quantities of honey. These are bees of the genus Apis and some of the Meliponinae (stingless bees).

Use of Honey

The main uses of honey are in pharmaceutical companies, cooking, baking, spreading on bread or toast, and as an addition to various beverages such as tea. Because honey is hygroscopic (drawing moisture from the air), a small quantity of honey added to a pastry recipe will retard staling. Raw honey also contains enzymes that help in its digestion, several vitamins and antioxidants. Honey is also used in traditional folk medicine and apitherapy, and is an excellent natural preservative.

Honey Products

Products which do not meet the compositional criteria for honey, but are products consisting in whole or in part of honey. Imitation or artificial honey is a mixture of sweeteners, colored and flavored to resemble honey. This product does not meet the definition of honey or honey products.

As such, it is inappropriate to include the word honey on the label of such a product. This is a partial and constantly growing list intended to standardize the vocabulary used in the honey trade.

1. **Deionized Honey**: A honey product where honey has been processed to remove selected ions.
2. **Deproteinized Honey**: A honey product from which protein has been removed by appropriate processing.
3. **Dried Honey**: Honey which has been dehydrated and in which edible drying aids and processing adjuncts may be included to facilitate processing and improve product stability.
4. **Honey Extract**: Any product formed by removing selected components from honey.The nature of the component (flavor, color, etc.) determines the type of extract.
5. **Honey Spread**: A variety of edible, extremely viscous honey products made from honey or creamed honey. Honey spread is sometimes blended with other ingredients (such as: fruits, nuts, flavors, spices or margarine but excluding refined sweeteners).
6. **Natural Honey Flavor**: A substance obtained (often by extraction) only from honey that contains the flavor constituents of honey.
7. **Ultra-filtered Sweetener Derived from Honey**: Honey from which all materials not passing a specified submicron membrane pore size have been removed. Materials removed include most proteins, enzymes and polypeptides. Evaporation required in the processing may also remove some volatile flavor and aroma constituents.

Honey Processing

Cut-Comb Honey

The simplest processing is to remove the honeycomb from frame hives, top-bar hives or traditional hives and sell or consume it as "cut-comb" honey. When producing this from frame hives it is necessary to use a wax foundation that does not contain strengthening wires and is thinner than that normally used in wired frames. The process involves collecting pieces of sealed and undamaged honeycomb, cutting them into uniform sized pieces and packaging them carefully in bags or cartons to avoid damaging the honeycomb. Because the honeycomb is unopened, it is readily seen to be pure, and it has a finer flavour than honey that is exposed to air or processed further. Cut-comb honey can therefore have a high local demand and fetch a higher price than processed honey. However, the honeycomb is easily damaged by handling and transport, which makes distribution for retail sale more difficult. It requires protection by packaging materials that will absorb shocks or vibration (e.g. cushioning plastics such as "bubble-wrap" and/or corrugated cardboard cartons) and packs should be carried carefully and not stacked, thrown or dropped to avoid damage to the honeycombs.

Strained Honey

This is honey that is processed to a minimal extent and is usually sold locally. It is prepared by removing the wax cappings of the honeycomb using a long sharp knife that has been heated by standing it in warm water. (un sealed combs containing unripe honey should not be used). The honeycombs are then broken into pieces and the honey is strained to remove wax and other debris. A fairly coarse strainer is used at first to remove large particles, and the honey is then strained through successively finer strainers such as cotton or muslin cloths. The clear honey is collected in a clean, dry container. When most of the honey has drained (often over many hours depending on the temperature) the combs are squeezed inside a cloth bag to remove as much of the remaining honey as possible. The wax is collected and formed into a block by melting it gently in a warm waterbath or solar wax extractor. This beeswax byproduct often has a high value as a wax polish or for candle-making. The strained honey can either be dispensed from the collection pan into customers's own containers or packed into glass jars or plastic bags for sale.

Processing Technology

Honey contains pollen, dust and air bubbles, which tend to include granulation (crystallization). Heating the honey to 45°C to dissolve the crystals present in honey can retard the granulation. Filtration then removes part of pollen, foreign particles and wax . To prevent fermentation and to destroy yeasts, honey is heated to a temperature of 65°C-70°C for specified time. Proper temperature and control and heating time is a most important factor in honey processing activity. Excessive heating increases the quality of Hydroxy- Meyhyl- furfal(HMF) which is desirable. High temperature also affects the color and flavor of honey. Honey is then cooled before it is packed to keep it for a longer period without contamination and granulation

1. Filtration to remove wax, foreign particles after heating honey to 45°C It may be noted that heating up to 45°C (below the melting point of bess wax) is required to decrease the viscosity of honey.
2. Honey is then heated to 60°C- 65°C for 10 to 15 min and passed in to a falling film evaporator. Vacuum is simultaneously applied to boil the water in honey at a lower temperature so that moisture is separated which can be collected separately. This procedure also helps in destroying yeasts.
3. Cooling the honey to atmospheric temperature and storing in closed vessel for 24-48 hrs. is the next step. Storing honey for period of 24-28 hrs. is necessary to allow air bubbles to go out . Honey is then packed and sealed immediately.

Table 31: List of machinery and equipment used for honey processing unit

S. No	Description
1	Liquifier
2	Filter press
3	Falling film Evaporator
4	Vacuum pump
5	Storage/settling tank
6	Water circulation pump
7	Pre heating tank
8	Processing tank
9	Cooling tank/condenser
10	Moisture conditioning tank
11	Honey circulation SS gear pump
12	Insulation (Optional)
13	Control panel, Level indicators, pressure gauges, temperature gauges, SS pipes and fittings.
14	Packing machines
15	Labeling machine
16	Working tables

Honey Production Process

Extraction

The honeycombs are inserted into an extractor, a large drum that employs centrifugal force to draw out the honey. Because the full combs can weigh as much as 5 lb (2.27 kg), the extractor is started at a slow speed to prevent the combs from breaking. As the extractor spins, the honey is pulled out and up against the walls. It drips down to the cone-shaped bottom and out of the extractor through a spigot. Positioned under the spigot is a honey bucket topped by two sieves, one coarse and one fine, to hold back wax particles and other debris. The honey is poured into drums and taken to the processing facility.

Processing/Refining

Modern processing facilities include heating and cooling units, filter presses and pumps that deliver the finished product to the packing line. Following are the key steps during refining process of honey:

- Liquification
- Pre-Heating and Straining

- Micro-filtration
- Inactivation of Yeast Cells (Processing)
- Vacuum Evaporation/moisture reduction
- Cooling of Honey

Filling

Lines include bottle cleaning, filling, capping, front and back labeling and packing. All finished goods are delivered to storage areas by a system of conveyors.

Labeling

Pre-printed front and back labels are pasted through machines to make certain that accurate names and sizes match up with every bottle turns out.

Warehousing

Finally, the finished products are stored in the warehouse and distributed according to the plan throughout the year as shelf life of honey is generally higher then other perishable items

Methods of Processing

Extracted Honey is honey only obtained by centrifuging decapped broodless combs.

Pressed Honey is honey obtained by pressing broodless combs with or without the application of moderate heat.

Drained Honey is honey obtained by draining decapped broodless combs.

Styles – honey which meets all the compositional and quality criteria of Essential composition & quality factors (mentioned in next section) of this standard may be presented as follows:

***(a) Honey**:* which is honey in liquid or crystalline state or a mixture of the two?

***(b) Comb Honey**:* which is honey stored by bees in the cells of freshly built broodless combs and which is sold in sealed whole combs or sections of such comb.

***(c) Chunk Honey**:* which is honey containing one or more pieces of comb honey;

***(d) Crystallized or Granulated Honey**:* which is honey that has undergone a natural process of solidification as a result of glucose crystallization;

***(e) Creamed (or Creamy or set) Honey**:* which has a fine crystalline structure and which may have undergone a physical process to give it that structure and to make it easy to spread,

By-Products / Waste

Four major byproducts of the honey-making process: beeswax, pollen, royal jelly, and propolis. Beeswax is produced in the bee's body as the nectar is transforming into honey. The bee expels the wax through glands in its abdomen. The colony uses the wax to cap the filled honeycomb cells. It is scrapped off the honeycomb by the beekeeper and can be sold to commercial manufacturers for use in the production of drugs, cosmetics, furniture polish, art materials, and candles.

Pollen sticks on the worker bee's legs as she collects flower nectar. Because pollen contains large amounts of vitamin B12 and vitamin E, and has a higher percentage of protein than beef, it is considered highly nutritious and is used as a dietary supplement. To collect it, the beekeeper will force the bees through a pollen trap an opening screened with five-mesh hardware cloth or a 0.1875-in (0.476-cm) diameter perforated metal plate. The single- or double-screened opening allows the pollen to drop from the bees' legs as they fly through. The pollen drops into a container and is immediately dried and stored.

Royal jelly is a creamy liquid produced and secreted by the nurse bees to feed the queen. Nutrient rich with proteins, amino acids, fatty acids, sugars, vitamins, and minerals, it is valued as a skin product and as a dietary supplement. Proponents believe it prolongs youthfulness by improving the skin, increases energy, and helps to reduce anxiety, sleeplessness, and memory loss.

Propolis is plant resin collected by the bees from the buds of plants and then mixed with enzymes, wax and pollen. Bees use it as a disinfectant, to cover cracks in the hive, and to decrease the hive opening during the winter months. Commercially it is used as a disinfectant, to treat corns, receding gums, and upper respiratory disease, and to varnish.

Honey Quality Control and Assurance

Honey is preserved because of its high sugar content (or conversely its low moisture content), which prevents micro-organisms (bacteria, yeasts and moulds) from growing in it. Despite this, it must be handled hygienically, and all equipment must be properly cleaned (see below). The aroma and taste of honey are its most important quality characteristics, but honey is often judged according to its colour. The colour of honey depends mainly on the source of the nectar. Usually dark-coloured honeys have a strong flavour whereas pale honeys have a more delicate flavour. Generally light-coloured honeys are more highly valued than dark products. Some honeys have a high pollen content, which makes them appear cloudy, and this may be considered as lower quality by some customers. The main causes of loss in quality of honey are:

1. An increase in moisture content - too much water in honey (greater than 19-20%) causes it to ferment. Honey is "hygroscopic", meaning that it will absorb moisture, and all honey processing equipment must therefore be completely

dry. Honey should also be processed as soon as possible after removal from the hive to prevent it absorbing moisture from the air, especially in humid climates. In areas with a very high humidity it can be difficult to produce honey of sufficiently low water content.

2. Development of HMF (Hydroxymethylfurfural). This is a break-down product of fructose (one of the main sugars in honey) that is formed slowly during storage but very quickly when honey is heated. Colour can also be an indicator of quality because honey becomes darker during storage and heating. The amount of HMF present in honey is used as a guide to the age of the honey and/or the amount of heating that has taken place. Some countries set an HMF limit for imported honey. HMF is measured by laboratory tests and technical advice from a Bureau of Standards should be sought if export is being considered.

3. Contamination by insects. Honey processing is a sticky operation, and the sugar in honey attracts ants, cockroaches and flying insects. Careful protection is needed at all stages of processing, including insect screens on doors and windows to prevent contamination by insects. All honey residues on equipment should be removed by proper cleaning to prevent them attracting insects. The presence of any other contaminations (e.g. particles of wax, parts of bees, splinters of wood, dust etc.) make the honey very low value.

Granulation

Glucose is one of the main sugars in honey and when it crystallises (i.e. it changes from a liquid to a solid), the liquid honey also becomes solid (or granulated). Depending on the source of the nectar collected by bees, some types of honey are more likely to granulate than others, but almost all honey will granulate if its temperature falls sufficiently. Granulation is a natural process and there is no difference in nutritional value between solid and liquid honey. Although there is obviously a difference in the texture between liquid and granulated honey, there is no difference in the flavour or other quality characteristics. Some customers prefer granulated honey, and if liquid honey is slow to granulate, the addition of 20% finely granulated honey will cause it to granulate.

Quality Checks

The routine quality checks on honey are a visual inspection to detect clarity, any contamination by insects or other materials such as particles of beeswax, and checking that the pack contains the correct weight of honey. In humid climates, or if a batch of honey is suspected of containing high a level of water (e.g. honey that is returned because it has started to ferment), it can be checked for moisture content. Because honey is mostly sugar (around 80%) and water (19-20%) the sugar content can be measured using a refractometer, and the value subtracted from 100 to measure the moisture content.

However, refractometers are expensive and it may be more affordable to send samples to a laboratory for checking if a problem with the moisture content is suspected. If during production, the level of moisture is too high, it can be reduced by blowing air for several hours over a pan of honey using an electric fan. Honey should never be heated to remove water because this will increase the amount of HMF and significantly reduce its quality. Cleaning The other important quality assurance check is to ensure that the correct cleaning procedure is in place and is being properly followed by production staff. All equipment, floors and work surfaces should be washed daily with hot water and detergent, and rinsed with clean water. They should be allowed to dry completely in the air before production starts again. Cloths should not be used to dry surfaces and equipment because they can contain sugar residues that recontaminate cleaned surfaces.

Table 32: Honey standards

S.No	Parameters	Standards
1	Fructose/Glucose Ratio	1.23
2	Fructose %	38.38
3	Glucose %	30.31
4	Minerals (Ash) %	0.169
5	Moisture %	17.2
6	Reducing sugars %	76.75
7	Sucrose %	1.31
8	pH	3.91
9	Total Acidity, mg/Kg	29.12

The analysis of the sugar content of honey is used for detecting adulteration.

Essential compositions and Quality factors of honey

A. Apparent Sucrose Content

1. Honey not listed below Not more than 5%
2. Honey dew honey, blends of honeydew honey and blossom Not more than 10%
3. Red bell (calothmnus sanguineus), white stringy bark (eucalyptus Scabra), Grand Banksia, & Blackboy Not more than 15%

B. Water Insoluble Solids Contents

1. Fornhoneys other than pressed honey Not more than 0.10%
2. Pressed Not more than 0.5%

C. Minéral Content (ash)

1. Honey not listed below Not more than 0.10%
2. Honeydew honey or a mixture of honeydew honey and blossom honey. Not more than 1.0%

D. Acidity

Not more than 40 milli equivalents acid per 1000 grams.

E. Hydroxy methyl furfural Content

Not more than 80 mg/Kg

Procedure

Preparation of Test Samples

Honey solution: 10.0 gram honey is weighted into a 50 ML beaker and 5.0 ML acetate buffer solution is added, together with 20 ML water to dissolve the sample. The sample is completely dissolved by stirring the cold solution. 3.0 ML sodium chloride solution is added to a 50ML volumetric flask and the dissolved honey sample is transferred to this and the volume adjusted to 50ML.

Note: It is essential that the honey should be buffered before coming into contact with sodium chloride.

Standardization of the Starch Solution

The starch solution is warmed to 40°C and 5 ML pipetted into 10ML of water at 40°C and mixed well. 1 ML of this solution is pipetted in to 10ML 0.0007 N iodine solution, diluted with 35 ML of water and mixed well. The color is read at 660 nm against a water blank using a 1 cm cell. The absorbance should be 0.760 ± 0.020. is necessary the volume of added water is adjusted to obtain the correct absorbance.

Absorbance Determination

Pipette 10 ML honey solution into 50 ML graduated cylinder and place in 40°C ± 2°C water bath with flask containing starch solution. After 15 minutes, pipette 5 ML starch solution into the honey solution mix, and start stop watch. At 5 minutes intervals remove 1 ML aliquots and add to 10.00 ML 0.0007 N iodine solutions. Mix and dilute to standard volume. Determine absorbance at 660 nm is spectrophotometer immediately using 1 cm cell. Continue taking 1 ML aliquots at intervals until absorbance of less than 0.235 is reached.

Jackfruit

Introduction

Jackfruit (*Artocarpus heterophyllus* Lamk.) is a tropical fruit species, grown in South East Asia regions, subjected to seasonal food shortages due to either drought or floods; hence it is popularly called as 'Poor man's fruit'. It is the national fruit of Bangladesh, indicating its importance in the economy of that country. Jackfruit is used in a variety of ways: fresh, canned and made into jam, candy, or chutney. The raw fruits are used as vegetable, pickles and as *papad*. The seeds are eaten roasted and or as flour. Jackfruit is rich in vitamins A and minerals, thus it is essential to local communities as sources of both nutritious foods and income.

Jackfruit is believed to be native of India, originated in Western Ghats and is widely cultivated in Bangladesh, Brazil, Malaysia, Myanmar, Pakistan, SriLanka, West Indies and other tropical countries. Jackfruit is the largest edible fruit in the world. It belongs to the family Moraceae and is a tropical evergreen tree.

Jackfruit grows naturally in all types of weather, and is extensively grown in Western Ghats region of South Karnataka, Kerala, TamilNadu, Andhra Pradesh, Maharastra, Andamon and Nicobar Islands, Assam, West Bengal and North Eastern part of India. The total area under Jackfruit in India, is approximately 1, 02, 000 ha (Bose *et al.*, 2003). In Karnataka, it is cultivated in an area of about 11,333 ha mostly in the Southern plains and Western ghats producing about 2.60 lakh tones of fruits per annum

Jackfruit is known for quality fruit, hardy wood, evergreen nature, producing copious organic matter and its ability to convert large quantity of CO_2 to oxygen. It adapts well to marginal conditions and its cultivation is simple, bears in five to six years. It is becoming popular as monocrop or pure crop to avoid the problem of labour and to get assured income from rainfed areas.

There are few sporadic efforts of value addition to Jackfruit on small scale in rural industries, whose efforts are laudable and needs to be encouraged to set up processing industries in a bigger scale, so that the produce in the local areas is absorbed and wastage is minimized with creation of jobs in the rural areas. Further, the unutilized portions of the fruit can be used in the preparation of animal feed and as organic manure.

Ripe Jackfruit is very nutritious and a rich source of pectin, carotene (250-1740 mg/100g) and minerals like phosphorous (97 mg/100 g), potassium (246 mg/100 g) and calcium (50mg/100g). Ripe jack bulbs are full of carbohydrates (19 g/100 g), protein, and ascorbic acid. Ripe bulbs are used in the preparation of canned products, nectar, jam, juice, fruit bar and candy. Unripe bulbs are used in the preparation of chips and papad.

The raw fruit is a popular vegetable in the west coast of Kerala, Karnataka, Tamil Nadu and is used in several culinary preparations. Due to its availability in plenty during the monsoon in the coastal regions and the non-availability of vegetables during that time has earned the name 'Poor Man's Food'. Foliage is a nutritive fodder and the wood has good timber value and is in great demand in many Asian countries due to its resistance to fungus and termites. Further, the seeds contain two lectins *viz.*, jackalin and artocarpin having diverse biological activities. With all these uses and economic benefits, the jackfruit is aptly called as " Kalpavruksha" or " wish granting tree".

Figure 40: Fresh Jackfruit pulp

Though Jackfruit is a native of India, it is widely grown in South and Southeast Asia *viz.,* Bangladesh, Indonesia, Pakistan, Philippines, SriLanka, Thailand and Vietnam. It is also grown in parts of central and eastern Africa, Brazil, Suriname and in islands of the West Indies such as Jamaica. It is a national fruit in Bangladesh.

In Philippines, Indonesia, Vietnam, Malaysia and Thailand, the Jackfruit has gained industrial status and they are preparing many value added products from both raw and ripe fruits for home consumption as well as for export purpose.

Why Process of Jackfruits?

- The fruit is rich in vitamins A,B and C, potassium, calcium, iron, protein and carbohydrates
- Processing reduce post harvest losses
- Processing increase the shelf life of the fruit
- Processing adds value and increase income

Figure 41: Jackfruit processing unit

Vietnam is the world leader in jackfruit products. It has 20 units that make chips. Among them Vinamit Trading

Figure 42: Whole and cut jackfruit

Corporation (Plate 1) is the biggest. In Vietnamese, mit means jackfruit. Vinamit exports jackfruit chips to countries like the US, Japan, Germany, Russia and China. Mr. Nguyen Lam Vien, CEO of Vinamit Company says that, for big land owners in Vietnam, growing rubber is profitable. But for small farmers, jackfruit fetches them more money than rubber.

Figure 43: Jackfruit in Jar

Profit from jackfruit is about $ 8,000 to $12,000 per hectare per year. The company has jackfruit orchards in about 10,000 hectares. It has entered into an eight year purchase contract with farmers. As a result, 300 new jackfruit orchards from five hectares to 100 hectares have been developed. They have selected elite jackfruit types, propagated vegetatively and distributed to farmers by the company itself (Civil Society- Aug. 2009- Online Farm Journal).

How to Store Jackfruits

- **Fresh fruits**: -for 4-5 days at 25-35°C

  -for 2-6 weeks at 11-13°C
- **Bulbs:** -for 3 weeks at 2°C, when packed in heat sealed polythene bags
- **Pulp:** -for more then one year at -20-22°C, when packed in heat sealed polythene bags.

Figure 44: Jackfruit chips package unit

or it may be too much to consume at once, particularly for smaller families. It is an aggregate fruit with numerous fruit lets/flakes, each containing one seed. The fruit let is covered with epidermal cells and cuticle layer with a waxy appearance. The process of separating the fruit lets from the centre core is quite unpleasant since the fruit is full of gummy latex that sticks to the hands and knives. The difficulty in accessing the flesh often results in an unsightly product, loosing consumer appeal.

Thus, the minimally processed jackfruit can be easily marketed on a daily basis or even can be stored frozen for export purpose. Polyethylene bag has been commonly used for packing minimally processed jackfruit for the local market and the stalls by the roadside. However, for export market, the minimally processed jackfruit can be packed in polystyrene tray overlapped with stretched film. Thus the storage life of minimally processed jackfruit was up to 3 weeks at 2 °C, 1 week at 10 °C and 2 days at 25 °C. The achievable storage life provides sufficient marketing planning for distribution both for local and export markets; which otherwise would have deteriorated rapidly, resulting in off-flavour. This technology of minimally processed jackfruit has a potential not only for the local markets but also for export.

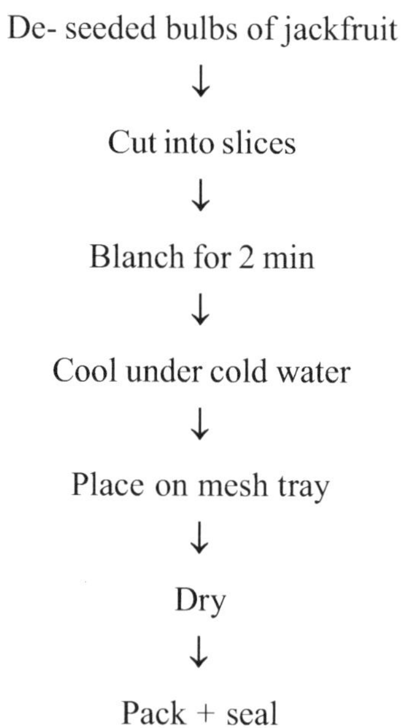

Figure 45: Flow diagram for Dries jackfruit flakes

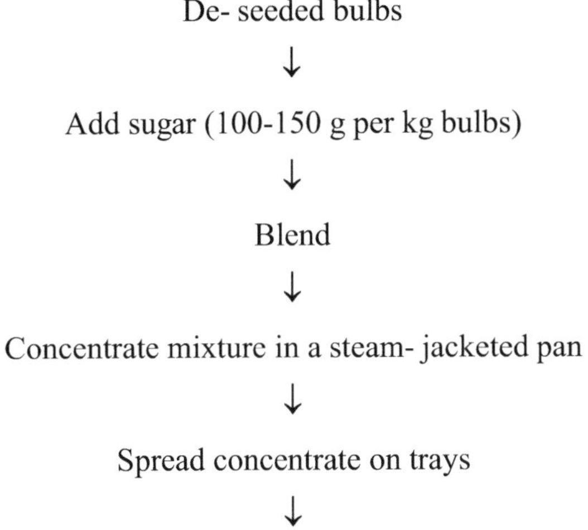

Dry+ Dust with starch→ cut into pieces and pack+seal

Figure 46: Flow diagram for Jackfruit Leather

Packaging and Storage of Dried Jackfruit Flakes

Pack in heat sealed 400gauge polythene bags and store in a dark, cool place for several months,

Packaging and Storage of Jackfruit Leather

Pack in heat sealed 400gauge polythene bags and store in a dark, cool place for several months,

References

Acedo Jr., A.L. 1992. Jackfruit Biology, Production, Use,and Philippine Research. Monograph Number 1. Forestry/Fuelwood Research and Development (F/FRED) Project,Arlington, Virginia.

Anonymous. 1998. Comoro Islands. Worldmark Encyclopedia of Nations. Africa. Gale, Detroit.

Campbell, R.J., and N. Ledesma. 2003. The Exotic Jackfruit: Growing the World's Largest Fruit. Fairchild Tropical Garden, Coral Gables, Florida.

Coronel, R.E. 1986. Promising Fruits of the Philippines. University of the Philippines at Los Baños, College of Agriculture, Laguna.

Crane, J.H., C.F. Balerdi, and R.J. Campbell. 2002. The Jackfruit (*Artocarpus heterophyllus* Lam.) in Florida. Fact Sheet HS-882. Horticultural Sciences Department, Florida Cooperative Extension Service, Institute of Food and Agricultural Sciences, University of Florida, Gainesville. http://edis.ifas.ufl.edu/BODY_MG370

Delang, Claudio O. 2006. The Role of Wild Food Plants in Poverty Alleviation and Biodiversity Conservation in Tropical Countries'.Progress in Development Studies 6(4): 275-286

Gale.2008, Forest Nurseries and Gathering of Forest Products. Encyclopedia of American Industries, 5th ed..

Gunasena, H.P.M. 1993. Documentary Survey on *Artocarpusheterophyllus* (Jackfruit) in Sri Lanka. Monograph Number 2. Forestry/Fuelwood Research and Development (F/FRED) Project, Winrock International, Arlington, Virginia.

Marla and Rebecca J. McLain; (editors). 2001. Non-Timber Forest Products: Medicinal Herbs, Fungi, Edible Fruits and Nuts, and Other Natural Products from the Forest. Food Products Press: Binghamton, New York.

http://www.agricultureinformation.com/forums/stevia/8089-medicinal-properties-stevia.html

http://www.ayurvedic-medicines.com/herbs/amla.html

http://www.formerfatguy.com/articles/stevia-medicinal.asp

http://www.for.gov.bc.ca/hfd/library/documents/glossary/Glossary.pdf. Retrieved 2009-04-06

http://www.hort.purdue.edu/newcrop/morton/bael_fruit.html

Guillen, Abraham; Laird, Sarah A.; Shanley, Patricia; Pierce, Alan R. (editors). 2002. Tapping the Green Market: Certification and Management of Non-Timber Forest Products. Earthscan

Jones, Eric T. Rebecca J. McLain, and James Weigand. eds. 2002. Non Timber Forest Products in the United States. Lawrence: University of Kansas Press.

Peters, Charles M.; Alwyn H. Gentry, Robert O. Mendelsohn (29). "Valuation of an Amazonian rainforest". Nature 339. doi:10.1038/339655a0.

Soepadmo, E. 1992. *Artocarpus heterophyllus* Lam. In: Verheij, E.W.M., and R.E. Coronel (eds.). Plant Resources of South East Asia 2. Edible Fruits and Nuts. Prosea, Bogor, Indonesia.

Wilkinson, K.M., and C.R. Elevitch. 2003. Propagation protocolfor production of container *Artocarpus heterophyllus*

Chapter - 6

Milk and Milk Products Based Food

Importance

The dairy products, which have originated in India, are called indigenous dairy products. The products can be broadly classified in to khoa based and channa based. The importance of the dairy products were known to Indians since time immemorial or it could be roughly estimated to be around five thousand years ago and the development could be considered as an art.

Table 33: Product lists for milk and milk products based convenience foods

S.No	Process	Products	Western Counterpart
1	Concentration	Khoa	Evaporated Milk
		Basundi	Sweetened Condensed Milk
		Rabri	
		Khurchan	
		Malai	
2	Coagulation	Channa	Cheese prepared by direct acidification
		Paneer	White cheese – Latin American origin
3	Fermentation	Dahi	Pain Yoghurt
		Payodhi	Sweetened Yoghurt
		Misti Dahi	
		Chakka	
		Shrikhand wadi	
		Lassi, Chaach	Stirred Yoghurt
4	Phase inversion of fat	Makkan	Butter
		Malai	
		Ghee	Butter oil
		Kulfi	Ice cream
		Malai ka baraf	

The term Indian Dairy Products refers to those milk products, which originated in undivided India

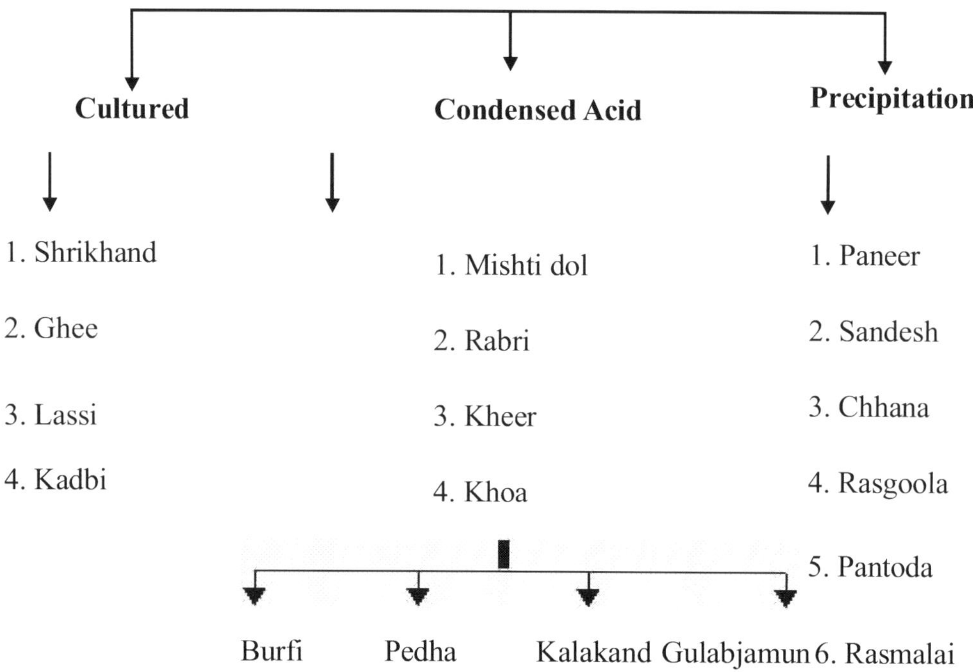

Figure 47: Flow chart of conversion of milk into traditional Indian dairy products

Cultured

Srikhand: - Srikhand is a semi-soft sweetish sour, whole milk product prepared form lactic fermented curd. The basic ingredient of Srikhand is Chakka.

Method of preparation: - the standardized method of preparation consist of fresh, sweet buffalo milk, which has been standardized to 6% fat, is pasteurized at 71^{0}C for 10 minutes and then cooled to 28-30^{0}C. It is then inoculated @ 1% with lactic culture which is mixed well, and incubated at 28-30^{0}C for 15-16 hours. When the curd has set firmly (acidity 0.7-0.8% lactic), it is broken and placed in a muslin cloth bag and removed after 8 to 10 hours. Now the curd gets change into a solid mass called Chakka. This Chakka is then mixed with grinded sugar. Colour and flavour can also be added to obtain the product known as Srikhand.

Ghee: - Ghee is a clarified butter fat prepared from cow or buffalo milk. The largest ghee producing states are U.P, A.P, Punjab, Rajasthan, M.P, Bihar, Hariyana etc. The production of ghee is higher in winter and lower in summer.

Method of preparation: - cream accumulated after few days is usually taken in a suitable vessel and heated and stirred on a low flame to remove the moisture contain. After removing moisture contain further heating is stopped then cooling is done. On cooling, when the residue has settled down the clear fat is decanted into suitable containers.

Table 34: Nutritional comparison between cow and buffalo milk

Characteristics	Cow	Buffalo
Milk fat	99 to 99.5%	
Moisture	Not more than 0.5%	
Unsaponiable matter		
Carotene (m g./g.)	3.2-7.4	-
Vit. A (I.U./g.)	19-34	17-38
Charred casein, salts of copperand iron, etc.	Max.2.8 (Agmark) Traces	

Lassi: - Lassi, also called chhas or matha, refers to desi buttermilk, which is the by-product obtained when churning curdled whole milk with crude indigenous devices for the production of desi butter (makkhan). It appears that 50-60kg. (ave.55kg) of lassi are produced for every kg of ghee. Lassi contains appreciable amounts of milk proteins and phospholipids.

Cheese: - Cheese has high protein content in a very digestible form, is rich in calcium and phosphorous, and is an excellent source of fat-soluble and water-soluble vitamins. It is concentrated form of energy, contributing 4cal/g and is also a highly suitable food for those suffering from lactose intolerance. Cheese is a bio-enriched food the enrichment being brought about by vitamins and micronutrients being produced as metabolites of the starter bacteria. Although the precise mechanisms are not known, there is sample evidence to suggest that the consumption of cheese at the end of a meal prevents dental caries by reducing the thickness of the film formed on teeth, the degree of abrasion of the chewing surfaces, dental surface defects and caries.

Table 35: Composition of Lassi

Characteristics	Milk (%)
Water	96.2
Total solids	3.8
Fat	0.8
Solids-not-fat	3.0
Protein	1.3
Lactose	1.2
Ash	0.4
Lactic acid	0.44

Traditional Indian Dairy Products

Condensed (Rabri)

It is prepared concentrated and sweetened product comprising of several layers of clotted cream. The layer of cream formed, as a skin is continuously removed. When the milk is reduced to 1/3 of the original volume, sugar is added and the layer of cream skin is mixed.

Method of Preparation

Rabri is normally prepared by heating 3-4kg of milk in a karahi over an fire to simmering temperature (85-90^{0}C), and then maintaining the temperature by controlled heating. The milk is neither stirred nor allowed to boil. The surface of the milk may be gently fanned to help the process of skin formation. A piece of this skin, about 3-4cm. square, is continuously broken with a thick wooden stick and moved to the cooler parts of the karahi. This operation requires considerable skill and constant attention. Simultaneously, as slow evaporation reduces the milk to about one-fifth of its original volume, good quality round sugar at 5-6 per cent by weight of the original milk is added to the milk concentrate and dissolved in it. The layers of skin collected on the karahi surface are then immersed in the mixture and the finished product obtained by gently heating the whole mass for another brief period.

Table 36: Composition of Rabri

Characteristics	Percentage
Moisture	30
Fat	20
Protein	10
Lactose	17
Ash	3
Sugar	20

Khoa: -khoa is a partially dehydrated whole milk product.

Method of preparation: -Milk is cautiously stirred in a circular motion to prevent scorching. When milk becomes viscous the rate of stirring is increased to maintain a uniform consistency. The pan is removed from the fire and the product is worked up with the flattened end of the scraper by alternatively spreading into thin layers and collecting repeatedly until it retains its shape. After cooling it becomes solid. Nearly 36% of the country's total khoa production takes place in U.P.

Table 37: Composition of khoa made from cow and buffalo milk

Characteristics	Cow	Buffalo
Moisture	25.6	19.2
Fat	25.7	37.1
Protein	19.2	17.8
Lactose	25.5	22.1
Ash	3.8	3.6
Iron (ppm)	103	101

Products of Khoa

A. Peda

Ingredients

Khoa (Mawa) 225g.

Sugar 75g.

Pista (optional) A few pieces

Silver paper (optional) 1 leaf

Cardamom (optional) A few sticks

Method

Break freshly made khoa (mawa) into bits. Mix (preferably ground) sugar into it. Put into a karahi and cook over a very slow non-smoky fire, stirring all the while with a khunti. Add crushed cardamom if desired. When mixture is ready (mixture forms balls when tested), pour into a tray and leave to cool and set. Peda is now ready. Decorate with sliced pista. Cut into required size and shape to serve.

Kalakand

Ingredients

Milk 1 kg

Sugar 60g.

Citric acid 1/2g.

Pista (optional) a few pieces

Silver paper (optional) 1 leaf

Cardamom A few sticks

Method

Boil the specific quantity of milk in a karahi placed over a brisk and non-smoky fire. Stir continuously with a khunti with a circular motion. After 10-15 minutes, add to it the required amount of citric acid as a dilute solution in water. These will partially coagulate the milk. At this time vigorous stirring is required to obtain a product of good quality. When a semi-solid stage is reached, add sugar and stir well. Add crushed cardamom if desired. Remove after five minutes. This finished product is set in a greasy tray or plate and allowed to cool at room temperature. Kalakand is now ready. Decorate, if desired, with silver paper and sliced pista. Cut into required size and shape to serve.

Gulabjamun

Ingredients

Khoa 300g.

Maida 35 g.

Baking powder ½ tsp. (teaspoonful)

Sugar 1kg.

Water 1kg.

Ghee 500g.

Method

Break the entire (freshly made) khoa into bits. Mix baking powder into the maida separately. Add this mixture to the broken khoa and mix again. Now start kneading by adding small quantities of water until uniform dough is obtained. While kneading, there should be no oozing of fat. To avoid this, especially in summer, keep the vessel in which the kneading is done upon a tray in which ice or chilled water is kept. The consistency of the dough should be such that when made into small balls it has a smooth uncreaked surface. Meanwhile dissolve all the sugar in water, and boil the solution till a 2-string-consistency-syrup is obtained.

During this process, and 4 tablespoonfuls of milk and ladle out the scum to obtain a clear syrup. Keep this in a container so that a minimum depth of about 10-cm of syrup is obtained. Now make the balls and test-fry in sufficient ghee or dalda (vegetable ghee) in a shallow karahi, so as to immerse the balls completely during frying. The balls should be neither over-nor under fired. They should be deep brown in colour. Cut one fried ball into two and examine the inside for porosity. If found satisfactory, then fry the whole lot. If insufficiently porous, add a minute quantity of baking soda solution, sprinkling and mixing it well into the dough, and repeat the process of test frying, etc. if it is too porous and the ball bursts when fried, add small lots of maida instead and repeat the testing process. Remove the balls and put them into syrup immediately, pressing them down in the sugar syrup for some time so that it soaks in. Keep gulabjamun at room temperature for at least 10-12 hours before serving.

Burfi

Ingredients

Khoa (fresh and hot) 250g.

Sugar (crystal) 75g.

Chocolate 10g.

Method

Break khoa into bits and spread it in a karahi. Add (preferably crystal) sugar to it and mix well by working vigorously with a wooden ladle. Collect the mixture into a compact mass when all the sugar has dissolved. This is Plain burfi. Now separate one third of the mixture and mix chocolate into it. Take a well-greased plate and spread plain burfi (two-thirds of the mixture) as a thick layer. Apply the chocolate-mixed portion all over it as a thin layer. Allow cooling and setting at room temperature. This is chocolate Burfi. Cut into desired size and shape to serve.

Kheer: Kheer is also known as Basundi. It is used for direct consumption as desert. It is prepared by concentrating milk to half of its original volume by open pan concentration and adding sugar and other condiments.

Method of preparation: - fresh, sweet, cleaned milk standardized to 4.0% fat and vigorously boiled in a jacketed stainless steel pan for 3 to 5 minutes accompanied by constant stirring cum scraping with a khunti. High-grade rice 2.5%of milk, pre-cleaned and washed with cold water before use, is now added. The mixture is gently boiled, with periodical stirring-cum-scraping. When the concentration is about 1:8:1, clean, good quality sugar is added @ 5% of milk. Gentle heating is continued for another 3 to 5 minutes till a final concentration of about 2:1 is obtained. The yield of finished kheer should be about 50% of the milk used.

Acid Precipitation

Pannier: - Pannier refers to the indigenous variety of rennet-coagulated, small-sized, soft cheese.

Table 38:Composition of Pannier made from cow and buffalo milk

Characteristics	Cow (%)	Buffalo (%)
Moisture	71.2	71.1
Fat	13.5	13.1
Total solids	28.8	28.9

Method of preparation: - Surati Cheese or pannier is the best known of the few indigenous varieties of cheese. The name Surati appears to have been derived from the town of Surat.

Technique of production: - Fresh buffalo milk, standardized to 6 per cent fat, is pasteurized by heating it to 78°C (172°F) for 20 seconds and promptly cooling it to 35°C (95°F). about 0.5 to 2 kg of this milk is placed in the coagulating pan and the temperature maintained at 35°C (95°F) by circulating warm water in the jacket. Good quality lactic starter @0.5 per cent of milk is now added to the milk and thoroughly mixed into it. This is followed by the addition of rennet @6-7ml/100 lit. milk, the rennet being previously diluted with about 20 times its volume of water (The quantity of rennet added should be such as to give a clean cut in the curd at the end of about 60 minutes.) after mixing it adequately, the renneted milk is allowed to set till a firm coagulum fit for basketing is obtained.

The temperature during this time is maintained at $35^0C/95^0F$. The curd is then ladled out with a vertical slant in thin slices, and filled into especially made bamboo/ wicker baskets. These baskets are previously prepared by cleaning them with heated water, keeping them soaked in a 10% lukewarm salt solution for about 10 minutes, and then thinly dressed with salt. Each successive layer of curd put into the baskets is uniformly sprinkled with salt. Salting is done @4-5% of the green cheese (which works out to approximately 2% of the milk taken).

After they have been filled, the baskets are placed on the draining rack to allow for drainage of whey, which is collected in a tray placed underneath. Generally, at the end of 50 to 60 minutes, the individual pieces of cheese are firm enough to be handled without breaking. At this stage, they are carefully turned upside down in their respective baskets. This is known as the 'First Turning'. After draining them for a further 30-40 minutes, the cheeses, on attaining the desired firmness and consistency, are subjected to their 'Second Turning'. The collected whey is then strained through a muslin cloth and kept in the cheese-soaking basin. The pieces of cheese are removed from the baskets and carefully submerged in the whey. They are then left steeped in whey for 12-36 hours till disposed of or used. The yield of surati panir is approximately 28.5% for cow and 34.0% for buffalo milk.

Chhana

Chhana, also called pannier in certain parts of the country, constitutes one of the two chief bases (the other being khoa) for the preparation of indigenous sweetmeats. Chhana refers to the milk-solids obtained by the acid coagulation of boiled hot whole milk and subsequent drainage of whey. The acids commonly used are lactic or citric, in both natural and chemical forms. It should not contain more than 70% moisture, and the milk fat content should not be less than 50.0 per cent of the dry matter.

Table 39: Composition of Chhana made from cow and buffalo milk

Characteristics	Cow (%)	Buffalo (%)
Moisture	53.4	51.6
Fat	24.8	29.6
Protein	17.4	14.4
Lactose	2.1	2.3
Ash	2.1	2.0

Method of production: - There are two methods for making chhana, which are adopted by commercial manufacturers.

Batch method: Usually all the milk for chhana-making is brought to boil by heating it directly in a large iron karahi over an open fire, all the while stirring it with a khunti, and later keeping it simmering hot in the karahi. This hot milk is ladled out in batches of 0.5 to 1kg into a separate coagulation vessel, either already containing, or to which is promptly added the required quantity of the coagulant. The latter is normally cleansed sour chhana-whey, which is maintained in a large earthen vessel from day to day. The mixture of milk and whey is stirred with the ladle, and when it has completely coagulated, the contents are poured over a piece of clean muslin cloth stretched over another vessel (for receiving the whey). The process is repeated till all the milk is used up. The cloth containing the coagulated solids is then removed, tied up into a bundle without applying pressure and hung up not only to drain out the whey completely but also to cool the chhana-pat.

Bulk method: All the milk (5-15kg) is brought to boil as above in the karahi, which is then removed from the fire. The coagulant is then added slowly and gradually in the required quantity to the entire lot of milk and stirred with the ladle so that it mixes properly and clear coagulation takes place. The chhana is collected by straining it through a cloth.

Sandesh

Ingredients

Chhana 250g.

Sugar 75g.

Flavour (optional) A few drops

Cardamom (optional) A few sticks

Method

Break freshly made chhana into bits. Mix (preferably ground) sugar into it. Put the mixture in a karahi and heat on a slow fire stirring all the time with a khunti. (Add crushed cardamom, if desired, towards the end). When the mixture is ready (mixture forms balls when tested) pour it into a tray and leave it to cool and set. Sandesh is now ready. It is cut or moulded into the desired size and shape. (A popular flavour-cum-colour is saffron, which is mixed with the finished product before it is cut or moulded).

Rossogolla

Ingredients

Chhana (soft) 200g.

Maida (optional) 8g.

Sugar 250g.

Water 1 kg.

Elaichidana A few pieces

Flavour (Rose) A few drops

Method

Break the above quantity above quantity of chhana into bits and start kneading. There should be no oozing of fat during this operation. To avoid this, especially during summer, keep the vessel in which the kneading is done, upon a tray in which ice or chilled water is kept. If required, a small quantity (as above) of maida may be added to avoid cracks in the finished rossogollas. The consistency of the kneaded mass should be such that when made into small balls, it has a smooth surface without signs of cracks. Meanwhile, dissolve all the sugar in water and boil the solution. During this process, add 2 tablespoonfuls of milk and ladle out the scum to obtain clear syrup. Keep this in a suitable-sized degchi in which the chhana balls are being cooked, such that a minimum depth of 10-15cm. of syrup is obtained.

Now make the balls of chhana. While doing this, one sugarcoated Elaichidana may be put in the centre of each ball. After all the balls (10-15) have been made, put them gently in the boiling sugar syrup for the cooking process. See that the balls do not overcrowd the degchi and that there is enough space for them to move freely, especially after they swell. Close the lid of the vessel. The heating should be so controlled that the balls are constantly covered with foam. Keep a watch from time to time. After 5-10 minutes, the balls will swell. If the chhana has been well made and properly kneaded, the balls will not crack or break. After 5-10 minutes, the colour of the balls will darken slightly. The finished rossogollas should normally be ready after 20-25 minutes. During the last stage, the lid should be removed so that the sugar syrup finally attains 1-string consistency. After cooling, sprinkle flavour (rose) to serve.

Kulfi: - Ice cream has frozen in small containers. While the milk is boiling, it is sweetened by an addition of sugar and the product is concentrated to approximately 2:1. To this concentrate, when it has cooled, are added malai (indigenous cream), crushed nuts and a flavour (commonly rose or vanilla). The mix is placed in triangular, conical or cylindrical moulds of various capacities made of galvanized iron sheets. The moulds are closed on top by placing a small disc over them. A mixture of ice and salt in the ratio of 1:1.

Commercial manufacture of whey based fruit drinks: Large quantity of soybean milk whey is produced during preparation of soy-*paneer* and it goes waste. Soybean milk whey based ready-to-serve (RTS) beverages were developed and evaluated. The properties of soybean milk whey were evaluated for its utilization for preparing RTS beverages.

Characteristics of soybean whey: Titratable acidity, protein and total reducing sugar of whey were observed to be 0.192, 0.3 and 0.4 % respectively. pH, average specific gravity and viscosity of the freshly prepared soybean whey were observed to be 5.11, 1.01 and 20.5 cP. Colour of the whey was yellowish red. The L*, a* and b* values of the whey were in the range of 32.02 – 40.35, 3.03 – 4.56 and 19.58 – 22.27 respectively. The turbidity of whey was 82.37 NTU.

Preparation of RTS beverages: RTS beverages were prepared by adding sugar, fruit juice / pulp, citric acid and preservatives to the whey. Different compositions were tried with guava, mango and pineapple. The soybean milk whey could be used upto 30 % by weight due to its typical aroma. pH, titratable acidity, TSS and sensory score of the RTS samples was observed to be in the range of 3.71 – 3.85, 0.25 – 0.37 %, 14.2 – 19.8 o Brix, 7.12 – 8.5 respectively. The microbial load of RTS samples after 15 days of storage was observed to be 1 – 2 x 106 /g fungal colonies and no bacterial contamination. The utilization of soybean milk whey for RTS beverages enhances the nutritional value of the beverages by providing the benefits of phyto-chemicals available in the whey.

Reference:

Arbuckle, W.S. 1986, Ice Cream. 4th ed. Chapman & Hall, London, New York.

Carci, M. 1994. Concentrated and Dried Dairy Products. VCH, New York.

DiLiello, L.R. 1982. Methods in Food and Dairy Microbiology. AVI, Westport, CT

Food and Agricultural Organization. 1990. The Technology of Traditional Milk Products in Developing Countries. Food and Agricultural Organization of the United Nations, Rome.

Eckles, C.H., W.B. Combs and H. Macy, Milk and Milk products, Tata McGraw-Hill, Bombay, 1980.

Fox, P.F.1992. Advanced Dairy Chemistry. Chapman & Hall, London, New York.

Ganguli, N.C.,1974. Milk Proteins, Indian Council of Agriculture Research, New Delhi, 1974.

Kurmann, J.A., Rasic, J.L., and Kroger, M. 1992. Encyclopedia of Fermented Fresh Milk Products: An International Inventory of Fermented Milk, Cream, Buttermilk, Whey, and Related Products. Chapman & Hall, London, New York.

Ranganna, K.S. and K.T. Achaya, Indian Dairy Products, Asia Publishing House, Bangalore, 1974

Warner, J.N., Principles of Dairy Processing, Wiley Eastern ltd., New Delhi, 1976

Robinson, R.K. 1994. Modern Dairy Technology. 2nd ed. Chapman & Hall, London, New York

Varnam, A.H. 1994. Milk and Milk Products: Technology, Chemistry and Microbiology. Chapman & Hall, London, New York.

Wong, N.P. 1988. Fundamentals of Dairy Chemistry. 3rd ed. Chapman & Hall, London, New York.

Chapter - 7

Food Additives used for Foods

Importance of Food Additives

Food additives are chemicals, biochemical ingredients etc., which are added to food products for their technological benefits. These benefits may be to preserve the quality of food, to maintain and improve its appeal, to ensure its nutritional value and to provide innovations in new products with consumer appeal and satisfaction. Additives are also added to maintain uniform quality and to enhance quality parameters such as flavour, colour, texture etc. in large-scale production. They and their by-products ultimately become part of the food that is consumed and so they must be safe.

Additives may be direct additives, which are added deliberately to improve its sensory quality, nutritive value, stability, ease in processing and retention of quality during handling and retailing. There are also indirect additives, which are not added intentionally but get included into foods incidentally during handling, processing and packaging.

There are some guiding principles for the use of food additives. They should be justified for their technological effectiveness and purpose. They should be safe for use. There should be maximum adequate levels, absolutely necessary levels of usage and ADI (Acceptable Daily Intake) properly evaluated while considering its safety and permitted usage levels in foods. They should not be added with the intention of misleading consumers about quality. They should also not significantly affect adversely the nutritional quality of food products.

Classification of Food Additives

- Preservatives
- Food Colours
- Food flavours and flavouring agents
- Emulsifying, Stabilizing, Anti caking and Antifoaming agents
- Antioxidants

- Sequestering and Buffering agents/ Acidulants
- High intensity / low calorie sweeteners
- Vitamins and minerals
- Processing Aids
- Nutraceuticals
- Probiotics/Prebiotics
- Functional additives

Coding of Food Additives

The **food additive coding system** was developed by the European Community (**EC**). The European food additive code numbers are prefixed by 'E' (e.g. E223). These E-numbers indicate the food additives that are approved for use in Europe.

Countries outside Europe use the numbers but do not add the E prefix. For example, acetic acid is written as E260 on food products sold in Europe, but it is known as additive 260 in Australia. Additive 103, alkanet, is approved for use in Australia and New Zealand, but is not approved for use in Europe and so does not have an E number.

Different types of food additives and their functions are listed in Table

Table 40: Classes and function of food additives

Class of additive	Function	Examples
Anti-caking agents	Keep powdered products (e.g. salt) flowing freely when poured	• Bentonite (558), • Calcium aluminium silicate (556), • Calcium silicate (552)
Anti-foaming agents	Reduce or prevent foaming in foods	• Polyethylene glycol 8000 (1521), • Triethyl citrate (1505)
Antioxidants	Retard or prevent the oxidative deterioration of foods	• Butylated hydroxyl anisole (320), • Ascorbyl palmitate (304), • Calcium ascorbate (302)
Artificial sweeteners	Impart a sweet taste for fewer kilojoules/calories than sugar	• Sorbitol (420), • Alitame (956), • Aspartame (951), • Saccharin / calcium saccharin (954)
Bleaching agents	Whiten foods	• Chlorine (925), • Chlorine dioxide (926), • Benzoyl peroxide (928)
Bulking agents	Increasing the bulk of a food without affecting its nutritional value	• Ammonium chloride (510), • Isomalt (953), • Polydextrose (1200)
Colourings	Add or restore colour to foods	• Curcumin (110), • Brilliant blue FCF (133), • Tartrazine (102)
Colour retention agents	Retain or intensify the colour of a food	• Ferrous gluconate (579)

Contd.

Emulsifiers	Prevent oil and water mixtures separating into layers	• Lecithin (322), • Sorbitan monostearate (491), • Ammonium salts of phosphatidic acids (442)
Enzymes	Break down foods (e.g. ferment milk into cheese)	• α-amylase (1100), • Lipases (1104), • Proteases (papain, bromelain, ficin) (1101)
Firming agents	Strengthen the structure of the food and prevent its collapse during processing	• Calcium chloride (509), • Calcium gluconate (578), • Calcium sulphate (516)
Flavour enhancers	Improve the flavour and/or aroma of a food	• Calcium glutamate (623), • Disodium 5'-ribonucleotides (635), • Ethyl maltol (637)
Food acids	Maintain a constant level of sourness in a food	• Acetic acid (260), • Citric acid (330), • Fumaric acid (297)
Flour treatment agents	Improve flour performance in bread making	• Sodium metabisulphite (223), • Ammonium chloride (510), • Potassium bromate (924)
Glazing agents	Impart a shiny appearance or provide a protective coating to a food	• Beeswax, white and yellow (901), • Carnauba wax (903), • Shellac (904)
Gelling agents	Thicken and stabilize various foods (e.g. jellies, deserts and candies)	• Agar (406), • Calcium alginate (404), • Carrageenan (407)
Humectants	Prevent foods from drying out (e.g. dried fruits)	• Glycerin or glycerol (422), • Lactitol (966), • Oxidised polyethylene (914)
Mineral salts	Improve the texture of a food (e.g. processed meats)	• Cupric sulphate (519)
Preservatives	Protect against deterioration caused by microorganisms	• Sodium nitrate (251), • Benzoic acid (210), • Sodium benzoate (211)
Propellants	Gases which help propel a food from a container	• Carbon dioxide (290), • Nitrogen (941), • Nitrous oxide (942)
Sequestrants	Bind and remove unwanted minerals that cause oxidation	• Potassium gluconate (577)
Stabilisers	Maintain the uniform dispersion of substances in a food	• Xanthum gum (415), • Guar gum (412), • Bleached starch (1403)
Thickeners	Improve texture and maintain uniform consistency	• Tannins (181), • Sodium alginate (401), • Pectins (440)
Vitamins	Restore vitamins lost in processing and storage	• B vitamins, including niacin • Vitamin C • Vitamin E

Differences Between Natural and Artificial Colouring

Artificial or synthetic colours are synthesised in the laboratory. They can be chemically identical to colours that occur naturally. Natural colours are derived from natural or biogenic sources (e.g. animal, vegetable or mineral). Both natural and artificial colours are used in food products like ice creams, confectionery, biscuits, sweet meats, fruit drinks, seasonings, pharmaceutical tablets and syrups.

The use of colour additives which might cause cancer or hyperactivity in children has raised concern among consumers. It is possible, but rare, to have an allergic-type reaction to a colour additive. For example, tartrazine (102) is an artificial colouring that has been associated with allergic reactions in some rare instances. Reactions have ranged from rashes and swelling to asthma, and possibly even to behavioural changes. Individuals should read food labels and avoid certain colour additives if they experience allergic reactions.

Food Additives Safe to Consume

Although safety assessments of food additives are carried out by PFA Act before the food additives are approved for use, food additives can still induce adverse reactions in some sensitive individuals. According to PFA Act, it does recognize the adverse reactions to food additives in a small proportion of the population. These reactions are not the same as allergies, but may include rashes and swelling of the skin, irritable bowel symptoms, behavioural changes in children, and headaches.

Two major groups of food sensitivity are known as food allergy and food intolerance. Food allergies are abnormal immunologic responses to a particular food or food component. In contrast, food intolerances are non-immunologic responses. Generally, total avoidance of the culprit food is necessary for true food allergies. Food intolerances can be managed by limiting the amount of the food or food ingredient that is eaten. Total avoidance is usually not necessary for food intolerances. Some commonly used food additives that tend to induce adverse reactions are:

Aspartame

Aspartame (E951) is an artificial sweetener that is used to replace sugars in foods and beverages. The long term effects of aspartame on health have been studied intensively, but results were inconclusive. It is noted that aspartame induces carcinogenic effects in a dose-related manner. Contradictory results were shown in studies which reported that aspartame consumption in foods and beverages does not raise the risk of brain or other cancers.

Although inconclusive results were shown in several studies, PFA Act and other international regulatory agencies concluded that aspartame is safe to consume. Aspartame is approved for general use in tabletop sweeteners, carbonated soft drinks, yoghurt and confectionery.

The acceptable daily intake (ADI) of aspartame is currently 50 mg/kg body weight in the United States and 40 mg/kg body weight in Australia and the European Union for both children and adults.

Benzoate

Sodium benzoate (E211) is used as a food colouring and preservative in foods. Children who consumed a mixture of food colourings and preservatives from soft drinks and confectionery at high levels were found to be more hyperactive than those who did not have the colourings and preservatives. Colourings and preservatives can be minimised in diets by including lots of fresh fruits and vegetables and eliminating processed foods.

Monosodium Glutamate (MSG)

Monosodium glutamate (E621) is often added to food as a flavour enhancer but it can also occur naturally in food. In the safety assessment conducted by FSANZ, MSG has been implicated as the causative agent of Chinese restaurant syndrome (CRS) and asthmatic attacks. However, CRS is only likely to occur when MSG is consumed in a large dose without food, and symptoms are not serious and may be attenuated when MSG is consumed with foods. MSG is not seen as the significant triggering factor of asthma in asthmatic individuals.

Nitrates

Nitrates or nitrites are added as a preservative, antimicrobial agent or colour fixative to processed foods such as meats and cheese. Nitrate also occurs naturally in water, vegetables and plants. The human body converts nitrate in food into nitrite. Nitrite has been implicated in a variety of long term health effects, including gastric cancer.

Sulphite

Sulphite sensitivity is a food intolerant reaction. Sulphites exist in several forms (e.g. sodium and potassium metabisulphite, sodium and potassium bisulphite, sodium sulphite, and sulfur dioxide). Sulphite has many functions, including as a antimicrobial agent. It inhibits enzymatic and nonenzymatic browning, whitens foods, and serves as a dough conditioner. Manifestations of sulphite sensitivity include anaphylaxis and asthma.

Tartrazine

Tartrazine (E102) is an approved artificial food colour. Tartrazine has been implicated in the aggravation of both asthma and chronic urticaria in some people. However, the association of tartrazine in the provocation of asthma and chronic urticaria is controversial. Some studies have shown a cause-and-effect relationship, whereas other studies have not. Both asthma and chronic urticaria are chronic illnesses with symptoms that tend to flare up at unpredictable times.

Sensitivity of Food Additives

Food additives used in food products are approved by PFA Act. However, some individuals are sensitive to specific food additives. The degree of sensitivity varies from person to person.

Children

There has been a debate regarding the detrimental effect of artificial colours and preservatives on the behaviour of children. It has been suggested that artificial food colours and other preservatives may produce overactive, impulsive and inattentive behaviours in children. However, the sensitivity of each individual to artificial colours is different. Some children have positive changes of behaviour when artificial colours are eliminated from their diet, while others do not. Children who have hyperactive or hyperkinetic behaviour are to a larger extent diagnosed with attention deficit hyperactivity disorder (ADHD).

Although ADHD is normally present at birth and tends to run in families, a study showed that food additives also predispose school-aged children to hyperactive behaviours. A child with ADHD has difficulty focusing his attention or engaging in quiet passive activities, and has educational difficulties, especially in relation to reading.

Asthmatic Effect:

Some food colours and preservatives (e.g. sulphite, tartrazine and MSG) can induce asthma. Therefore, asthmatic individuals should beware of these components in food products and beverages.

Looking for Additives in Foods

Food additives can be found in the **ingredients list** on the **food label.** Food labeling enables consumers to identify the presence of additives in packaged food and to make an informed choice about the foods they buy.

According to PFA Act, food additives are required to be identified by their class name, followed by an individual name or code number. To simplify the food labels, **a code number** is used to replace the individual name of the food additives. Food additives may be listed in food labels as, for example, thickener (guar gum) or thickener (412).

Consumer Concerns and Issues of Food Additives

Consumers today are quite aware of certain issues concerning food processing and additives are concerned about some aspects with respect to additives. Of primary importance is the issue of safety. Some additives have been banned because of their safety problems. This worries consumers and they are skeptical about many additives. They sometimes wonder whether there are health hazards associated of any of the additives over prolonged usage. Some media reports and activists who write about ill effects of some additives. This makes them more concerned about the safety issues.

Quality is another concern regarding use of additives. There have been reports in the past of use of some additives to hide poor quality. They feel that additives are added to misguide them about the quality of food products. Although this may be possible in some low quality products, most of the organized industry uses additives to improve the sensory quality but not to misguide consumers in that respect. However, this doubt gives rise to another issue and that is 'need to know'. They would like label declaration about additives so they can make a choice, between kinds of additives or without any additives.

Since there are additives that are natural and synthetic, and consumers feel natural are safe, these concerns regarding synthetic additives is also quite significant. Among other concerns are packaging and the value for the money spent on packaged food products. Branded products have made a definite impact on consumers and many are willing to spend extra for branded, packaged food products although some may not. Another big concern about packaged foods is the claims on the labels. Some misleading and fraudulent claims in the past have created a negative image about claims. Consumers constantly are on guard about the claims.

Food Additives: Approval Process

Any new additive before approving must undergo rigorous toxicity studies, including acute and chronic studies involving biochemical evaluation, teratogenic studies, and reproductive studies besides the LD_{50} tests. In the US, Delaney Clause governs the approval of any food additive, under which the additive is banned if found to be carcinogenic, under any condition or level, a very difficult zero risk condition. However, exposure assessment is very important in determining the risk involving any additive under the modern practice of determining safety. The Risk Analysis, adopted nowadays involves, risk assessment, wherein the Hazard is identified & characterized, Exposure is assessed and thus risk is characterized. Once the Risk is assessed, it must then be managed so hazardous conditions do not arise. Finally the risk must then be communicated.

References

Clydesale, F. (ed), Food Science and Nutrition, Prentice- Hall, Inc., New Jersey, 1979

Fennema, O.R., Food Chemistry, (Chap. 10) Marcell Dekker Inc., New York, 1976

Furia, T.E., Hand Book of Food Additives, Vol. I.C.R.C. Press, T.C.Ohio, 1977.

Hanssen M, Marsden J. The new additive code breaker: Everything you should know about additives in your food: Complete number guide for Australia and New Zealand. 2nd ed. Port Melbourne: Lothian; 1991.

Food Standard Australia New Zealand. Food additives [online]. 2007 [cited 2008 June 26]. Available from url: http://www.foodstandards.gov.au/ foodmatters/foodadditives.cfm

New Zealand Food Safety Authority. Identifying Food Additives [Booklet]. 2008 [cited 2008 July 3]. Available from url: http://www.nzfsa.govt.nz/ consumers/ chemicals-toxins-additives/ additives-booklet.pdf

Food Standard Australia New Zealand. Food additives: Alphabetical list [online]. 2007 [cited 2008 June 26].Available from url: http://www.foodstandards.gov.au/ newsroom/ publications/ choosingtherightstuff/ foodadditivesalphaup1679.cfm

Food Standard Australia New Zealand. Aspartame [online]. 2003 [cited 2008 June 27]. Available from: http://www.foodstandards.gov.au/ newsroom/ factsheets/ factsheets 2007/aspartame-september203703.cfm

Chapter - 8

Food Package Development

Definition

Packaging is the science, art, and technology of enclosing or protecting products for distribution, storage, sale, and use. Packaging also refers to the process of design, evaluation, and production of packages. Food packaging is packaging for food. It requires protection, tampering resistance, and special physical, chemical, or biological needs. It also shows the product that is labeled to show any nutrition information on the food being consumed

Packaging may be defined as co-ordinated system of preparing foods for transport, distribution, storage, retailing and their use. It is also a means of ensuring safe delivery of goods to the consumer in sound condition without affecting adversely the quality of the product packed in it at a reasonable cost. Packaging thus has as "techno-economic function" aimed at maintaining the quality of food stuff packed, with a view to retain the quality for a reasonable period.

Packaging plays a vital and dominant role in marketing, in maintaining quality and in preserving the food.

Consequent to the increase in the consumer awareness of the healthy hazards, which may be posed by different types of packaging material. It becomes inevitable for the packers to use better quality, hygienic packaging material which could retain the quality of the product for a longer time without posing any health hazard. The choice of a suitable packaging becomes all the more necessary for a number of factors like physico-chemical characters and acceptability of the commodities packed in any packaging material. Apart from this the package should be of the right shape and size and must attract the consumer's eye which could offer an absolute confidence regarding the wholesomeness of the products pack in it.

Objectives of Packaging

- **Physical protection** – The objects enclosed in the package may require protection from, among other things, mechanical shock, vibration, electrostatic discharge, compression, temperature, etc.

- **Barrier protection** – A barrier from oxygen, water vapor, dust, etc., is often required. Permeation is a critical factor in design. Some packages contain desiccants or Oxygen absorbers to help extend shelf life. Modified atmospheres or controlled atmospheres are also maintained in some food packages. Keeping the contents clean, fresh, sterile and safe for the intended shelf life is a primary function.
- **Containment or agglomeration** – Small objects are typically grouped together in one package for reasons of efficiency. For example, a single box of 1000 pencils requires less physical handling than 1000 single pencils. Liquids, powders, and granular materials need containment.
- **Information transmission** – Packages and labels communicate how to use, transport, recycle, or dispose of the package or product. With pharmaceuticals, food, medical, and chemical products, some types of information are required by governments. Some packages and labels also are used for track and trace purposes.
- **Marketing** – The packaging and labels can be used by marketers to encourage potential buyers to purchase the product. Package graphic design and physical design have been important and constantly evolving phenomenon for several decades. Marketing communications and graphic design are applied to the surface of the package and (in many cases) the point of sale display.
- **Security** – Packaging can play an important role in reducing the security risks of shipment. Packages can be made with improved tamper resistance to deter tampering and also can have tamper-evident features to help indicate tampering. Packages can be engineered to help reduce the risks of package pilferage: Some package constructions are more resistant to pilferage and some have pilfer indicating seals. Packages may include authentication seals and use security printing to help indicate that the package and contents are not counterfeit. Packages also can include anti-theft devices, such as dye-packs, RFID tags, or electronic article surveillance tags that can be activated or detected by devices at exit points and require specialized tools to deactivate. Using packaging in this way is a means of loss prevention.
- **Convenience** – Packages can have features that add convenience in distribution, handling, stacking, display, sale, opening, reclosing, use, dispensing, reuse, recycling, and ease of disposal
- **Portion control** – Single serving or single dosage packaging has a precise amount of contents to control usage. Bulk commodities (such as salt) can be divided into packages that are a more suitable size for individual households. It is also aids the control of inventory: selling sealed one-liter-bottles of milk, rather than having people bring their own bottles to fill themselves.

Method of Packaging

Metal containers, glass containers, rigid and flexible packaging materials are available for packaging food materials. Among these flexible packaging materials are commonly used for packaging convenience foods.

Flexible Packaging Materials

A. Retort Pouches

Retort pouches are flexible packages made from multilayer plastic films with or without aluminium foil as one of a layer. Their most important feature is that unlike usual flexible pouches they are made of heat-resistant plastics, thus making them suitable for processing in retort at temperatures of around 121°C normally encountered in thermal sterilization of foods. Besides, retort pouches should possess toughness and puncture resistance normally required for any flexible packaging to be machinable in pouch making and packaging operations and also withstand the rigours of handling and distribution. Apart from being heat sealable, the pouch material should possess good barrier properties to give the desired shelf-life to the product and be suitable for food contact applications

The 3-ply laminate consisting of PET/Al foil / PP is the most common material used in retort pouches and is the only one used in India at present. PET or polyester, used as the outer layer in thickness of about 12 mm, gives the required strength to the pouch. PET with its excellent glossy surface and printability, can be reverse printed so that inks are embedded between the outer layer and the next inner layer to give striking graphics. This property is presently not being used in the country, since the pouch is again packed in a paperboard carton. Aluminium foil used in the thickness of 9 to 25 mm serves as the barrier layer which is responsible for the shelf-life of more than one year obtained for the product. Polypropylene (PP) used as the inner layer usually in the thickness of about 75 mm provides the critical seal integrity, flexibility, strength and taste and odour compatibility with a variety of food products.

Other materials commonly used in retort pouch structures include nylon, silica – coated nylon, ethylene vinyl alcohol (EVOH) and polyvinyledaene chloride (PVDC). Nylon has a limited gas barrier properties and in combination with PP, it can be used for products that do not require more than3-6 months shelf-life.

With the more recent availability of high gas barrier films such as EVOH PVDC and silica coated nylon, non foil based retort-pouches that give long shelf-life have also become popular.

B. Aseptic Packaging

Aseptic packaging materials for unit packs and bulk packaging have been developed for packaging of ready to serve fruit beverages and fruit pulps / concentrates respectively.

The material used for unit packs are made of polyethylene / paper board / PE/ foil / PE / PE (tetra pack) (ii) PE/ foil / PE (iii) MET PET / PE (iv) composite cans (paper / foil/ PE). For bulk packaging, bag in box and bag in drum are made of met. PET/PE or PET/ Foil / PE.

PE - Polyethylene

PET - Polyethylene terephthalate

Type of Food Packaging

Drink boxes, an example of aseptic processing, the above materials are fashioned into different types of food packages and containers such as:

Packaging type	Type of container	Food examples
Aseptic processing		
Primary	Liquid whole eggs	
Plastic trays	Primary	Portion of fish
Bags	Primary	Potato chips
Boxes	Secondary	Box of Cola
Cans	Primary	Can of Tomato soup.
Cartons	Primary	Carton of eggs
Flexible packaging	Primary	Bagged salad
Pallets	Tertiary	A series of boxes on a single pallet used to transport from the manufacturing plant to a distribution center.
Wrappers	Tertiary	Used to wrap the boxes on the pallet for transport.

Primary packaging is the main package that holds the food that is being processed. Secondary packaging combines the primary packages into one box being made. Tertiary packaging combines all of the secondary packages into one pallet.

There are also special containers that combine different technologies for maximum durability:

- Bags-In-Boxes (used for soft drink syrup, other liquid products, and meat products)
- Wine box (used for wine)

Package Materials Suitable for Foods

1. Fried and puffed dhal: Fried and puffed dhal are quite crisp in nature. These are prone to pick up moisture when stored under humid conditions. Storage studies conducted in different packaging materials such as polyethylene and polyester / polyethylene films and paper at foil polyethylene revealed that oxidative off flavours are the major cause of storage deterioration through increase in peroxide value did not correlate the acceptability of stored products.

Fried green gram dhal was found to remain stable for 120 days at 37°C and for 210 days at 27°C when packed in paper aluminium foil polyethylene laminate pouches. In polyethylene and polyester / polyethylene films the shelf life was considerably shorter.

2. Extruded pellets: These are double extruded products, prepared from wheat, rice, maize flour or from their blends. Immediately after extrusion the moisture content in pellets is around 20-25%, which is brought to 7-8%, by hot air drying in a tunnel dryer.

The dried pellets are packed in attractive moisture proof packs of plastic films. These pellets equilibrate to 0.45 – 0.60 and remain stable for more than a months at room temperature. Before consumption these have to be fried in oil which imparts crispness and softness and about 4-10 times expansion in volume.

One of the important advantages of pellet type products is that they have long shelf life and any off flavour generated during storage gets volatilized during frying operation. Secondly the product is not fragile and packaging requirements are relatively less stringent compared to ready to eat expanded products.

3. Instant mixes based on chemical leavening: Instant mixes such as gulabjamun, cake, pancake, ice cream, dosai, idli mixes belong to this category. The formulated mix is invariable packed in polyethylene pouches which are encased in printed duplex card board packets control of granularity, moisture and adequate mixing are essential for product acceptability and shelf life.

4. Instant mixes based on precooked dehydrated products: Important among this group are instant pulav, khichadi, bisibalebhat, curried dal, rice, peas, curried chholay, sambar, rasam, dahir, uppuma mixes. These products are packed in polyethylene or poly propylene pouches, while defence specifications require hermetically sealed polypropylene pouch to be further encased in heat sealed paper aluminium foil polyethylene laminate pouch. These have to be finally packed in 5-ply corrugated cartons followed by polyethylene film over wrapping to protect from moisture and insects.

5. Ready to eat products stabilized by antimycotic agent: Chapathi, parota and poori are most popular, traditional wheat products consumed all over India. Freshly baked chapathies are soft and pliable but on storage they become tough and brittle due to moisture loss and starch retrogradation. After 24-48 hours these lend to develop fermented odours, ropiness and mould growth.

Extensive studies undertaken mainly by DFRL to preserve these products ready to eat state for development of operational pack rations for armed forces. Attempts to preserve chapaties by incorporation of propionates were less successful but by incorporating 0.4%, sorbic acid, packing in moisture proof laminate packs and preserved for six months under ambient conditions. Chapathies preserved by this method were found to be acceptable to troops and mountainers upto 6 months storage.

Bread, buns, cakes are other convenience marketed in ready to eat state. Though both sorbates and propionates are permitted under PFA, only propionates are used by baking industry in bread and buns to achieve a shelf life of around 3 days. On the other hand, wrapping bread in fungi static wrapper having 2 gm sorbic acid per sq. metre has been claimed to extend the shelf life upto 10 days. In cakes sorbic acid is permitted to higher concentration (0.15%) which ensures considerably longer shelf life under normal trade channels.

6. Canned convenience foods: Some attempts have been made to preserve some of the traditional Indian foods like upma, halwa, parothas, subji, dal, curries etc. idli, kheer, pulav, peas, paneer curry etc. by thermal processing in cans mainly for supply to troops. Among these canned uppma and kheer are being supplied regularly for Antartica expeditious. Both these items have a shelf of around one year.

7. Other items: Instant rava idli mix has a shelf life of 8 months at room temperature and 6 months at 37°C when packed in PET pouches. Suji Halwa was prepared and packed in polypropylene (PP) pouches (8 x 12 cms) and hermetically sealed using an impulse heat sealer. The sealed pouches were heated in a hot air drying cabinet at 95°C for 1 and 2 hrs respectively. Samples packed in PP and PEP pouches were stored at room temperature (15-35°C), in refrigerator (2-5°C) and deep freezer (-12°C). Results showed that a refrigerated storage, halwa packed in polypropylene packs remained stable for 7 days with oil and for 20, 15 and 12 days respectively with added potassium sorbate (0.25%), calcium propionate (0.5%) and sodium benzoate (0.25%). At 20°C in a deep freeze, halwa packed in PP and PEP packs remained in highly acceptable conditions for more than 6 months without any perceptible changes in colour and flavour.

Packaging of Ready- to- Eat/ Ready-to- Cook Food

Convenience food is a concept that is prevalent in the developed world since long, while its inception into the Indian market has been recent. With the changing socio-economic pattern of life and the increasing number of working couples, the concept is fast becoming popular in Indian market. This type of food is becoming popular because it saves time and labour. This food has extended shelf-life and is available off the market shelves.

Packaged/convenience food products sector has been slow in penetrating the large potential presented by Indian 250 million strong middle class. But due to growing urbanisation and changing food habits, the demand has been rising at a good pace and there is enough latent market potential waiting to be exploited through developmental efforts.

The convenience food could be basically classified into two categories:

- Shelf – stable convenience food
- Frozen convenience food

Shelf-stable convenience foods are further classified as:

- Ready-to-Eat (RTE) and Ready-To-Serve (RTS) food - e.g. Idlis, dosas, pav bhaji, meat products like pre-cooked sausages, ham, chicken products, curries, chapattis, rice, vegetables like aloo chole, navratan kurma, channa masala etc.

- Ready-to-Cook food – e.g. instant mixes like cake mixes, gulab-jamun mix, falooda mix, icecream mix, jelly mix, pudding mix etc., pasta products like noodles, macaroni, vermicelli etc.

Packaging Requirements of Ready-to-Eat (RTE)/Ready-to-Serve (RTS) Food

A "ready-to-eat" food product may be defined as any food product which does not require any elaborate processing procedures on the part of consumer before it is good enough for consumption. It is ready-to-eat as soon as the pack is opened in a form, which is tasty and appetizing.

The advancements in food technology and packaging technology have made it possible to extend the shelf-life of these products. Before deciding which packaging material is to be used, it is necessary to know the packaging requirements of the product i.e. what hazards will cause product deterioration and the conditions to which the packaged product will be subjected throughout its shelf-life. Some important packaging considerations, which influence the selection criteria for choosing packaging materials, are highlighted.

- Product Characteristics
  - ❖ The type of food and its composition, moisture, fat, protein, flavour etc.
  - ❖ Form and shape of the product – smooth, regular, irregular, with sharp edges etc.
  - ❖ Nature of the product – crisp, brittle, sticky etc.
- Factors Affecting Packaging

  Factors responsible for the spoilage of the food products:
  - ❖ Biological spoilage due to micro-organisms
  - ❖ Abiotic spoilage due to chemical reactions like oxidation, hydrolysis and enzymatic reactions.
  - ❖ The environmental factors like light, humidity and temperature.
  - ❖ The food processing parameters eg. processing temperature and duration.
  - ❖ The shelf-life desired for a given ready-to-eat food, influences the type of packaging and processing parameters to be used.

Ready-to-eat snacks like idlis, dosas, pav bhaji etc. are sold across the counter and have a very short shelf-life, hence the packaging requirements of these products are different from those of ready to eat products like curry rice, upma, vegetable biryani etc., which are retort processed for longer shelf-life.

Products like idlis, dosas, pizzas are packed in packaging materials having low water vapour and oxygen permeability, odour and grease resistance, and good physical strength. The packaging materials generally used are injection moulded plastic containers, plastic film/bag pouches or paperboard cartons. In normal practice, the ready-to-eat food are consumed in a short span of time, but with the advancement in packaging technology, it is now possible to produce these items commercially and to extend the shelf-life up to a few years. Ready-to-Eat Products Packed in Retort Packs

Table 41: Use of Various Packaging Laminates/Composites

Material	Properties	Use
9 mm foil / adhesive /paper coated with heat sealing vinyl resin	Good moisture barrier, runs well on machine	Over wraps confectioneries
9 mm foil/adhesive/paper polyethylene (extruded)	Good moisture barrierruns well on machine	Fin-sealed pouches and sachets Over wraps for soups, etc.
1 in. polyethylene / 9 mm foil adhesive / paper	Heat seals by the wax bleeding through the tissue	Over wraps for confectioneries
9 mm foil / adhesive / paper / micro wax comp /tissue (20 g/m^2)	Low WVTR	Over wraps for biscuits, etc
Foil	Excellent WVTR, good machinability	Candy wrap, biscuit wrap
Cellophane / wad / cellophane	Excellent WVTR, good standwichprinting, good machine performance	Bags or pouches forhygroscopic items
Cellophane / adhesive / pliofilm	Excellent gas barrier, transparent pack	Nut packing with inert gas
Cellophane/polyethylene	Excellent gas barrier, trapped printing	Chocolate, etc.
Polyester film / Saran coated polyethylene	High strength, positive sealing	Vacuum food Pouches
Polyester / adhesive / foil / polyethylene	Excellent gas barrier good heat resistance, good rigidity, aroma retention	Flexible processable cans

Indian food like palak paneer, dal fry, curry rice, upma, vegetables biryani etc. are retort processed hence their packaging requirements are different. These products are retort processed because they are low acid food with moderate to large size particles; hence it is easy to remove oxygen from the head-space by gas flushing. The selection of a polymer or its combination is based on the requirement of barrier properties.

Retort pouch is a special package in which the perishable food items are preserved by physical, and/or chemical means. It is a flexible laminate, which can withstand thermal processing, and combines the advantages of the metal can and the boil-inbag. Ready-to-use retort pouches are flexible packages made from multilayer plastic films with or without aluminium foil as one of the layers. Unlike the usual flexible packages, they are made of heat resistant plastics, thus making them suitable for processing in retort at a temperature of around 121°C. These retort pouches posses toughness and puncture resistance normally required for any flexible packaging. It can also withstand the rigours of handling and distribution. The material is heat sealable and has good barrier properties.

In India, 3-ply laminate consisting of PET/Al Foil/PP is commonly used for packaging of ready to eat retort packed food. The product packed in such laminates has a shelf-life of one year. The other materials generally used in retort pouch structure includes nylon, silica coated nylon, ethylene vinyl alcohol (EVOH) and polyvinyledene chloride (PVDC). These materials have high moisture barrier properties and are used successfully for packaging of ready-to-eat high moisture Indian food.

Both preformed pouches as well as pouches formed on FFS machines are used. Preformed pouches are of flat and stand - up type. The typical structure of these pouches are:

- Flat configuration : 12μ PET/12μ Al foil/ 75μ PP
- Stand up configuration : 12μ PET/9μ Al foil/ 15μ OPA/60μ PP

The pouches are printed in attractive colours. The retort pouch is a space saving package by value of its design. It is a good substitute for tinplate cans as it eliminates the need for the addition of brine in the food. In conclusion, the market for retort pouches is certainly one that will continue to experience growth over the next few years, as the retort pouch gets acceptable as equal to or even superior to glass or metal containers. The pouch has the same shelf-life as the can or the Ready-to-Cook Products Packed in Flexible Plastic Pouches jar. The retort pouch needs to address ease of opening and re-closing (compared to glass jars). In addition, the packaging economics of the pack for mass volume products will depend upon the ability to increase filling speeds and to move from batch to continuous retort processing. Also, the retort pouch can save about 60% energy while processing. Furthermore, as the product is already sterile, it does not require additional low temperature storage.

Packaging Requirements of Ready-to-Cook (RTC) Food

Based on their initial moisture content, RTC food can be broadly classified as:

Low Moisture Food

- Moisture 1 to 5%
- Equilibrium Relative Humidity (ERH) 18-20%

These food have very low moisture and ERH. Hence they have the tendency to absorb moisture from the surroundings and turn soggy, thereby, loosing their crisp, brittle nature and taste. The most important factor to be considered is moisture vapour transmission rate (MVTR) of the packaging materials used. MVTR values of less than 1 gm / m2 / 24 hours are required.

Medium Moisture Food

- Moisture 6 to 20%
- ERH up to 65%
- Typical examples: Indian savory snacks, sweetmeats

Barrier property (MVTR) requirement for these food is less stringent, however, for longer shelf-life, microbiological spoilage has to be given due importance. Use of preservatives is often required.

High Moisture Food

- Moisture 20 to 60%
- ERH up to 85%
- Typical examples: Freshly baked products – bread, cake, chapatti, pickles, chutneys, sauces etc. Vermiclli in Plastic Pouches Plastic Pouches for Ready-to-Cook Food

For freshly baked products such as bread, cake, ERH is often higher than the ambient ERH.

Therefore the products tend to breathe out the moisture and if excess water vapour is not allowed to escape from the closed package, condensation on the outer surface of the product occurs, spoiling the product quality and leading to mould / yeast growth. Plastic films such as low density polyethylene (LDPE), which are permeable to water vapour are normally used for packaging these products for shorter shelf-life. For longer shelf-life, microbial spoilage is the main consideration. The products are sterilized and packed in hermetically sealed containers such as cans, retort pouches or aseptic packs. Medium and high moisture food are very susceptible to the microbial spoilage and need adequate processing and preservation methods, prior to their packaging.

Oxygen /Air Permeability

RTC food normally contain fat as well as other ingredients that can be oxidised. If oxygen/ air is allowed to come in contact with the packaged food, oxidative degradation of fat occurs, and many other oxidative changes take place, which cause rancidity, off flavor and discolouration in the food. Hence, packaging material for high fat should have low oxygen permeability.

Nitrogen Permeability

To protect the food from oxygen/moisture, the food is usually packed in an inert atmosphere of Nitrogen (N2). The N2 permeability of the package should be low to prevent its escape into the atmosphere.

Grease Resistance Properties

A variety of RTC food have edible oil and fat as their ingredients. Fat/oil during storage should not adversely affect the packaging material used for these products, as fat may ooze out.

HDPE and LDPE are affected by fat and are not suitable for packing fatty products. Polyester films, cellophane, polypropylene, inomer films etc. are suitable for such applications. If made in laminates, then the film offering excellent grease resistance is used as the innermost liner of the laminate. Flavour and essential oils contribute to the organoleptic qualities of many RTC food. They are volatile substances and hence gas permeability of the packaging material should be very low to prevent flavour loss. This is also necessary to block the entry of the outside oxygen and air, which could bring out the oxidative changes in flavour.

Light Sensitivity

Light accelerates oxidative changes associated with the flavours and fats in food. Opaque packaging materials such as cans and aluminium foil offer best protection from light. Metallised polyester and pigmented plastics are found quite satisfactory. Light could cause discolouration in coloured food. Some films are opaque to visible light but allow U.V. light to penetrate.

Based on their major ingredients the ready to cook mixes can be divided into four groups:

- Cereal based
- Legumes based
- Fat rich, and
- Spice enriched mixes

The first category consisting of mixes for idli, dosa, chakli are mainly sensitive to moisture pick up only and require protection against this. These generally have moisture content in the region of 8 to 10% and become soft and unacceptable at about 12 to 13% moisture content. Polyolefin plastic pouches of 37 to 75μ thickness are generally used for packaging, which provides 3-4 months shelf-life.

Legume or pulses based mixes comprise vada, khara sev, bonda, urad bath etc. have packaging requirements similar to those of cereal based mixes, but have lower permissible moisture pickup. Hence, this requires packaging material having good water vapour impermeability. LDPE and PP pouches have been found to offer 1½ to 3 months storage life under high temperature/RH conditions, which would correspond to 5 to 6 months under normal conditions. Ready mixes of Jamun, cake, doughnut etc., which have high fat content and milk solids are susceptible to rancidity and interactions with oxygen and water vapour. CPP pouches of 200 grams capacity have been found to give short shelf-life of 2-3 months, which is adequate for local marketing. However, for longer shelf-life and export purposes, plain printed polyester with LDPE or HD-LDPE co-extruded films would be better suited from the point of protection and attractive appearance.

Spice enriched mixes such as those of rasam, sambar, soup, bisibele bath are highly susceptible to aroma loss, oxidative deterioration changes and seepage of oil. Unsupported PE or PP pouches are inadequate to pack these items. More functional ones based on cellophane/PE, plain or metallised PET/PE, and co-extruded films with polyamide core layer provide longer shelf-life. Food mixes such as orange peel gravy, tamarind sauce etc. have very low moisture pickup tolerance and necessitate the use of highly fat resistant and flavour resistant packages. Innermost layer of HD-LDPE coextruded film, ethylene-acrylic acid copolymer provide the required properties and good heat sealability.

Frozen Convenience Food

The current trend in frozen food is dual ovenability i.e. products that can be heated in a microwave oven or a conventional oven. Shelf – stable retortable food are better suited for microwave heating. Aluminium trays which represented 85% of the market in the eighties are being replaced by other materials like paperboard, thermoset plastics and thermoplastics owing to the growing importance of microwave ovens. Among the three materials, paperboard has a image as well as functional problem. Consumers perceive it as a low quality material. Also, it softens in the presence of moisture and chars under high temperature conditions. Thermoset plastics also have several disadvantages. It is expensive and heavy, which increases shipping costs. It is brittle, stains easily and processes slowly.

Therefore, processors are looking to other materials like thermoplastics. Three critical properties to be considered when selecting thermoplastics for dual-ovenable packages are dimensional stability up to 200°C to 250°C, good impact strength at freezer temperatures to reduce shipping and storage damage and microwaveability. Other important properties include compliance with FDA regulations, absence of taste and odour, good release characteristics so that the food does not stick. This property is very important for baked food. The different types of dual ovenable packages are explained here.

Ovenable Plastic Based Food Trays

These trays are manufactured by thermoforming sheets of polypropylene (PP), high impact polystyrene (HIPS) and crystalline polyethylene terephthalate (CPET), each material offering specific advantages in performance and economics. The trays are vacuum formed or thermoformed from a reel of sheet. When extended shelf-life is required, PP is co-extruded with barrier resins such as EVOH to improve barrier properties for forming. PP trays cannot withstand conventional oven temperatures and are used only for microwave ovens. Foamed polystyrene trays with special blends of low density polystyrene can withstand much higher temperatures, however, they are used generally only for microwave with an advantage of good cost saving as compared to the CPET trays. CPET trays have distinct advantage of dual ovenability. They also very remarkably withstand the abuse of retail distribution. Their other advantages include design flexibility, resistance to oil and grease and no appreciable effect on food taste. CPET trays are stable from – 40°C to as high as 200°C and exhibit improved oxygen and water vapour barrier. All plastic trays are topped with heat-sealable lidding films or snap-on plastic domes.

Ovenable Board

Earlier developments were based on paperboard coated with TPX. This was expensive and the preferred material now is solid bleached sulphate board, extrusion coated with polyethylene terephthalate (PET). This material is resistant to exposure in hot-air ovens and to temperatures of 200-250°C. It is also used for containers for food to be re-heated in microwave ovens only.

The coated board is made into containers by two main methods. One technique produces containers by press forming to give trays or dishes similar to the conventionally pressed foil trays. An alternative system is based on existing carton technology and erects trays from flat carton blanks. The main reasons for the current interest in ovenable board trays are:

- Growing popularity of convenience food
- Developments of the microwave oven
- Developments of polyester coated ovenable board that is suitable for use in both microwave and conventional hot - air ovens

Ovenable board containers have to satisfy a number of performance requirements. First and foremost, the material must be permeable to microwave radiation. Metal surface reflects microwave radiation so that the aluminium foil dishes are not really suitable for microwave oven use. Containers intended for general use must also be heat resistant at temperatures up to 200-250°C, which will normally be encountered in hot air ovens. Resistance to heat includes a requirement that there should be no thermal degradation, browning or odour development. The material in contact with the food must be chemically inert and have food contact approval. It should also be grease resistant. The coating should be heat sealable and the material as a whole should be easily convertible at high speeds. Because the filled containers will normally be stored under deep-freeze conditions, the ovenable board must have good deep-freeze performance. Good printability is also a requirement.

Production of PET-coated board is carried out by extrusion coating. Pre-treatment of the board is necessary to give good adhesion of the coating. The behaviour of the total coated structure of both flame and corona board is limited by the cohesion of the clay coating.

References

American Society for Testing and Materials. 1991. Selected ASTM Standards on Packaging. 3rd ed. American Society for Testing and Materials, Philadelphia, PA.

Bakker, M. (Editor-in-Chief). 1986. The Wiley Encyclopedia of Package Technology. John Wiley & Sons, New York.

Brown, W.E. 1992. Plastics in Food Packaging: Properties, Design, and Fabrication. With Contributions by C.F. Finch, A. Speigel, and J.H. Heckman, Marcel Dekker, New York.

Hirsch, A. 1991. Flexible Food Packaging : Questions and Answers. Chapman & Hall, London, New York.

Holdsworth, S.D. 1992. Aseptic Processing and Packaging of Food Products. Chapman & Hall, London.

Jenkins, W.A. and Harringotn, J.P. 1991.Packaging Foods with Plastics. Technomic Publishing Co. Lancaster, PA.

Leonard, E.A. 1987. Packaging : Specifications, Purchasing, and Quality Control. 3rd ed. Marcel Dekker, New York.

Paine, F.A. and Paine, H.Y. 1992. A Handbook of Food Packaging. 2nd ed. Chapman & Hall, London, New York.

Chapter - 9

Food Storage and Quality Control

Importance of Food Storage

Food storage is both a traditional domestic skill and is important industrially. Food is stored by almost every human society and by many animals. Storing of food has several main purposes:

- Storage of harvested and processed plant and animal food products for distribution to consumers
- Enabling a better balanced diet throughout the year
- Reducing kitchen waste by preserving unused or uneaten food for later use
- Preserving pantry food, such as spices or dry ingredients like rice and flour, for eventual use in cooking
- Preparedness for catastrophes, emergencies and periods of food scarcity or famine
- Religious reasons (Example: LDS Church leaders instruct church members to store food)
- Protection from animals or theft

The guidelines vary for safe storage of vegetables under dry conditions (without refrigerating or freezing). This is because different vegetables have different characteristics, for example, tomatoes contain a lot of water, while root vegetables such as carrots and potatoes contain less. These factors, and many others, affect the amount of time that a vegetable can be kept in dry storage, as well as the temperature needed to preserve its usefulness.

The following guideline shows the required dry storage conditions

- Cool and dry: onions
- Cool and moist: root vegetables, potatoes, cabbage
- Warm and dry: winter squash, pumpkins, sweet potatoes, dried hot peppers

All vegetables can be safely stored by refrigeration or freezing. When refrigerated uncooked, they should be used within a month, although depending upon the extent of ripeness when refrigerated, they may spoil earlier; this will most likely be obvious due to visible mold, brown spots, and so on. Frozen vegetables should be used within a year.

When cooked, vegetables can be safely refrigerated for several days, or frozen for up to a year. However, if meat or animal ingredients such as fats or butter have been added to the vegetables during cooking, they should be eaten within 4 days if they are refrigerated; they can be safely frozen for 6 months.

Many cultures have developed innovative ways of preserving vegetables so that they can be stored for several months between harvest seasons. Techniques include pickling, home canning, food dehydration, or storage in a root cellar.

Storage Methods

Fruit

Fruits can be refrigerated for up to a month uncooked, or frozen for up to a year. There are a large number of methods of preserving fruit for extended periods of up to six months: candying, dehydrating, or cooking into preserves such as jam or jelly.

Grain

Grain, which includes dry kitchen ingredients such as flour, rice, millet, couscous, cornmeal, and so on, can be stored in rigid sealed containers to prevent moisture contamination or insect or rodent infestation. For kitchen use, glass containers are the most traditional method. During the 20th century plastic containers were introduced for kitchen use. They are now sold in a vast variety of sizes and designs.

Metal cans are used (in the USA the smallest practical grain storage uses closed-top #10 metal cans). Storage in grain sacks is ineffective; mold and pests destroy a 25 kg cloth sack of grain in a year, even if stored off the ground in a dry area. On the ground or damp concrete, grain can spoil in as little as three days, and the grain might have to be dried before it can be milled. Food stored under unsuitable conditions should not be purchased or used because of risk of spoilage. To test whether grain is still good, sprout some. If it sprouts, it is still good, but if not, it should not be eaten. It may take up to a week for grains to sprout. When in doubt about the safety of the food, throw it out.

Spices and Herbs

Spices and herbs are today often sold prepackaged in a way that is convenient for pantry storage. The packaging has a dual purpose of both storing and dispensing the spices or herbs. They are sold in small glass or plastic containers, or resealable plastic packaging. When spices or herbs are homegrown or bought in bulk, they can be stored at home in glass or plastic containers. They can be stored for extended periods, in some cases for years. However, after 6 months to a year, spices and herbs will gradually lose their flavour as oils they contain will slowly evaporate during storage.

Spices and herbs can be preserved in vinegar for short periods of up to a month, creating a flavoured vinegar.

Alternative methods for preserving herbs include freezing in water or unsalted butter. Herbs can be chopped and added to water in an ice cube tray. After freezing, the ice cubes are emptied into a plastic freezer bag for storing in the freezer. Herbs also can be stirred into a bowl with unsalted butter, then spread on wax paper and rolled into a cylinder shape. The wax paper roll containing the butter and herbs is then stored in a freezer, and can be cut off in the desired amount for cooking. Using either of these techniques, the herbs should be used within a year.

Meat

Perishable meats should be refrigerated or frozen promptly. Dry aging techniques are sometimes used to tenderize specialty gourmet meats by hanging them in carefully controlled environments for up to 21 days. Semi-dried meats like salamis and country style hams are processed first with salt, smoke, sugar, or acid, or other "cures" then hung in cool dry storage for extended periods, sometimes exceeding a year.

Unpreserved meat has only a relatively short life in storage. Pork and chicken should be eaten within one day, but beef and venison improve with up to 5 days storage in a cold room. When refrigerated, meats should be eaten by the "Best before" date stamped on the packaging, but not after that date.

Fish and Shellfish

It is unsafe to store fish or shellfish without preservation. Fresh fish and shellfish should be eaten within a few hours after harvesting, although it can safely be kept refrigerated for several days for selling. To store fish or shellfish for an extended period of time, it must be stored using methods such as freezing, drying (dehydrating) or pickling.

Food Rotation

Food rotation is important to preserve freshness. When food is rotated, the food that has been in storage the longest is used first. As food is used, new food is added to the pantry to replace it; the essential rationale is to use the oldest food as soon as possible so that nothing is in storage too long and becomes unsafe to eat. Labelling food with paper labels on the storage container, marking the date that the container is placed in storage, can make this practise simpler. The best way to rotate food storage is to prepare meals with stored food on a daily basis.

Commercial Food Storage

Silos connected to a grain elevator on a farm in Israel. Grain and beans are stored in tall grain elevators, almost always at a rail head near the point of production. The grain is shipped to a final user in hopper cars. In the former Soviet Union, where harvest was poorly controlled, grain was often irradiated at the point of production to suppress mold and insects. In the U.S., threshing and drying is performed in the field, and transport is nearly sterile and in large containers that effectively suppresses pest access, which eliminates the need for irradiation. At any given time, the U.S. usually has about two weeks worth of stored grains for the population.

Fresh fruits and vegetables are sometimes packed in plastic packages and cups for fresh premium markets, or placed in large plastic tubs for sauce and soup processors. Fruits and vegetables are usually refrigerated at the earliest possible moment, and even so have a shelf life of two weeks or less.

In the United States, livestock is usually transported live, slaughtered at a major distribution point, hung and transported for two days to a week in refrigerated rail cars, and then butchered and sold locally. Before refrigerated rail cars, meat had to be transported live, and this placed its cost so high that only farmers and the wealthy could afford it every day. In Europe much meat is transported live and slaughtered close to the point of sale. In much of Africa and Asia most meat is for local populations is raised, slaughtered and eaten locally, which is believed to be much less stressful for the animals involved and minimizes meat storage needs. In Australia and New Zealand, where a large proportion of meat production is for export, meat is stored in very large freezer plants before being shipped overseas in freezer ships.

Emergency Preparation

Guides for surviving emergency conditions in many parts of the world recommend maintaining a store of essential foods; typically water, cereals, oil, dried milk, and protein rich foods such as beans, lentils, tinned meat and fish. A basic food storage calculator can be used to help determine how much of these staple foods a person would need to store in order to sustain life for one full year. In addition to storing the basic food items many people choose to supplement their food storage with frozen or preserved garden-grown fruits and vegetables and freeze-dried or canned produce. An unvarying diet of basic staple foods prepared in the same manner can cause appetite exhaustion, leading to less caloric intake. Another benefit to having a basic supply of food storage in the home is for the cost savings. Costs of dry bulk foods (before preparation) are often considerably less than convenience and fresh foods purchased at local markets or supermarkets. There is a significant market in convenience foods for campers, such as dehydrated food products.

The Measures to Prevent Insect Infestation

1. **Storage facility:** Darkness and dampness would provide better condition for insect multiplication. Cracks and crevices in the walls and gaps in the floors when present, it favours insect infestation. To prevent this air-tight storage facilities is required. The storage area should be fumigated before keeping / storing the bulk materials.
2. **Moisture content of the foods:** The foods containing lower than 11% moisture are relatively resistant to insect attack. Proper heat treatment is required before packing the convenience foods.
3. **Packaging materials:** The convenience foods should be packed in air tight containers or moisture proof pouches.
4. **Infra-red radiation :** Low dose (50-100 K rads) inpack irradation is helpful to prevent insect infestation.
5. **Use of fumigants:** such as EDB, malathion and phosphine.
6. Incidence of insect infestation and measures to prevent infestation

Improper packaging and also faulty pretreatments during processing of convenience foods results insect infestations during storage. During the preparation of convenience foods, improper heat treatment to the raw ingredients results existence of moisture in the final product. Moisture proof packaging materials are necessary for packing convenience foods because most of the convenience foods especially the milk based and fruit based convenience foods are hygroscopic and pick up moisture easily. The moisture content exceeds the safer level results insect infestation. The common insects found in convenience foods are khapra beetle (*Trogoderma granarium*), Stegobium paniceum (Drug store beetle) red flour beetle (*Tribolium castaneum*) and Sitophilus oryzae. Stegobium paniceum is mostly attacked / infested in the spicy convenience foods (masala mixes and soup mixes).

Insect infestation brings about a reduction in the vitamins of 'B' group and denaturation of protein, palatability and digestibility of the insect damaged food is affected when damage has been caused upto a certain level. Food infested by insects cannot be freed of insect fragments and excreta of insects and these could be capable to causing serious damage to the health of the consumer. Heating and spoilage brought about by the developing insects in the grain, result in damage for greater than that caused by feeding. Presence of an appreciable proportion of weeviled grain in any of the convenience food will reduce its values for end – uses.

Quality Evaluation of Foods

Storage Behavior Physical and Chemical Evaluation

The quality of food is the basic criterion that determines the acceptability of any food product. The quality of the food is evaluated in two ways one is subjective evaluation and the other one is objective evaluation.

1. Subjective evaluation is sensory evaluation of prepared and also raw foods. Sensory evaluation is a multidisciplinary science that uses human panelists and their sense of sight, smell, taste, tough and hearing to measure the sensory characteristics and acceptability of food products. Thus, quality of food is judged in terms of appearance, odour, taste, texture and sound.

2. Objective evaluation include physical and chemical test. The following physical tests were conducted to study the storage behaviour of the prepared convenient food mixes.

Bulk density: The prepared product was weighed accurately. The volume of the weighed product was measured by mustard replacement method. The weight by volume was expressed as bulk density. This method is used for the identification of bulk density of idli, vadai, noodles, baji and other products.

Porosity test: In the case of idli porosity test was used to find out texture. Idli was cut into two at the centre and the ink blot test was done to study the texture of idli by observing the porosity.

Cooking time, cooked volume, cooked weight water, absorption and rehydration ratio was used for the evaluation of noodles, vermicelli sphagethi and macaroni.

Weight, diameter and thickness were evaluated for the papads, vadagam and kakra.

The frying characteristic was also studied for the evaluation of papads and vadagam. It was studied by deep fat frying in refined oil at 195°C for five to seven seconds. The important characteristics viz., oil absorption, expansion in diameter and blister formation was observed.

Physical characteristics: In the case of ice cream mix, the physical characteristics such as yield, volume, freezing temperature, freezing time, surface, texture, crystal formation, melting time and viscosity was studied.

Yield: The quantity of ice cream prepared from 100 gm of instant ice cream mix was noted.

Volume: The volumes of the ice cream were noted before and after freezing and the values were expressed in ml/gram.

Freezing temperature: The freezing temperature of the product was observed by using thermometer. The temperature was maintained at –20°C for conducting the study.

Freezing time: The time taken for the sample to freeze was noted and expressed in hours / minutes / seconds.

Surface: To determine the best quality of the product, appearance is taken important criteria. The surface of the product should be flat, smooth and not appeared as bulged.

Texture: The texture of the product was estimated by scooping. It should be soft, smooth and creamy in texture.

Crystal formation: The presence of ice crystal in the product was noted when the ice cream was eaten.

Melting time: The time taken for the product to defrost completely was noted and expressed in minutes / seconds.

Viscosity: The prepared ice cream sample was melted and allowed to flow through a tube of constant length and the flow rate was noted and expressed as ml per seconds.

Breaking strength: This test was used for the extruded products. An extruded product strip of 6.0 cm length was taken and its ends were tied horizontally to supporting strands about 15 cm above the ground level. A known weight (5, 10 and 15 gm) was dropped on the tied strip at different height, when the extruded strip breaks the breaking strength of the sample was recorded x gm for y cm.

Rehydration Characteristics of the Extruded Product

$$\text{Rehydration ratio} \quad \frac{\text{Drained weight of the rehydrated sample}}{\text{Weight of the dehydrated sample}}$$

Chemical analysis

1.	Moisture	–	estimated by hot air oven method
2.	TSS	–	estimated by using brix hydrometer
3.	pH	–	estimated by using pH meter
4.	Acidity	–	estimated by using titrimetric method
5.	Sugars (reducing	–	Shaffer somogyi method and total sugars)
6.	Starch	–	by anthrone method
7.	Amylose	–	by colorimetric method
8.	Protein	–	by micro kjeldhal method
9.	Fat	–	by solvent extraction method
10.	Free fatty acid	–	by tritrimetric method
11.	Peroxide value	–	by tritrimetric method
12.	Thiobarbituric acid	–	by tritrimetric method
13.	Ash	–	by using muffle furnace
14.	Minerals – calcium	–	by titrimetric method
15.	Mineral - iron	–	by calorimetric method
16.	Mineral – phosphorus	–	by calorimetric method

Microbial Load of the Foods

The hygienic quality of a food product is judged by microbiological analysis for indicator organisms which showed the safety and keeping quality by revealing the sanitation at the place of manufacture and also during processing, marketing, transportation and storage.

Geetha (2000) studied the microbial load of laboratory made ice cream mix and market samples. The microbial population was noticed in both instant and prepared ice creams of market and laboratory made samples.

Instant Mixes

The market samples had higher bacterial count than laboratory made. The bacterial count of market samples ranged from 3.6 to 5.4 x 10^3 cfu per gm whereas the laboratory made samples contained 1 x 10^3 cfu per gm of bacterial count. The fungi counts of market samples were 2 x 10^{-3}. The laboratory made had 1 x 10^{-3} cfu per gm of fungal population. The increase in bacterial and fungal counts of market samples might be due to improper handling of the samples during processing.

Prepared Ice Cream

The bacterial count of ice creams prepared from market instant ice cream mixes exhibited higher bacterial count than the laboratory made. The bacterial count of prepared ice cream of market samples recorded between 7.2 and 12.5 x 10^{-3} cfu per gm. The laboratory made samples showed 5.0-6.2 x 10^{-3} cfu per g of bacterial count. The ice cream prepared from instant ice cream mixes did not show any variations in the fungal population in both laboratory made and market samples with a few exceptions.

Karwasra *et al.* 2000 reported that the average bacterial count in different brands of ice cream samples ranged between 5.768 ± 0.225 and 1.968 ± 0.169 (log/g).

Sampatha *et al.* (1981) found that the canned carrot halwa was free from yeast mold and also from thermophilic anaerobes and E.coli.

Premavalli *et al.* (1991) reported that the vegetable halwas (carrot and pumpkin) without any treatment developed fermented odours within 10 days of storage. They also noted that the total yeast count was high in the control samples studied.

Sixty two samples of different types of instant mixed masala powders collected from Nagpur market were analysed for microbial content by Saxena *et al.* (1998). From the study it had been observed that 17 samples have exceeded the suggested limits of SPC by ICMSF. It was unusual to note that the count of **Staphylococcus** species was not observed in majority of mixed masala powders (46 out of 62 samples) and even the highest count noticed was 6 x 10^3 cfu/g. The mould count was much with in the limits lowest in Jaljira.

Thirty samples of instant channa masala and thirty five samples of meat masala of various brands were collected from different markets of North India. Channa masala and meat masala had microbial assay showed that total plate counts (TPC), aerobic thermophiles, aerobic spore formers were within the permissible limits, yeast / mould counts were not noticed in any of the samples (Kalra *et al.,* 1998).

Microbiological load of the 62 gulab jamum mixes was estimated by Rita Israni *et al.,* (1999). Yeast was not observed in any of the samples studied while moulds and Staphylococci was observed in 29 and 22%, of the samples respectively. Coliforms were noted in all except 2 samples taken for study.

References

Board, R.G., A. Modern Introduction to Food Microbiology, Blackwell Scientific Publications Oxford, 1983.

Food and Drink Manufacture, 1989. Good Manufacturing practice: A Guide to its Responsible Management. 2nd ed. Institute of Food Science and Technology, London.

Fung, D.Y.C. and Matthews, R.F. (Editors). 1991. Instrumental Methods for Quality Assurance in Foods, Marcel Dekker, New York

Gould, W.A. 1992. Total Quality Management for the Food Industries. CTC Publications, Inc., Baltimore, MD.

Harrigan, W.F. and Park, R.W.A. 1991. Making Safe Food: A Management Guide for Microbiological Quality, Academic Press, London.

Hayes, G.D. 1991. Quality in the Food Industry: A Manager's Guide to Current Developments and Future Trends. Technical Communications/G.D. Hayes, Manchester.

Herschdoerfer, Ed., 1984. Quality Control in the Food Industry. 2nd ed. Academic Press, London,

Hubbard, W.T., Hagstad, H,V.,and Spangler, E. 1991. Food Safety & Quality Assurance: Foods of Animal Origin. Iowa State University Press, Ames, IA

Kramer, A. and Twigg, B.A. 1970. Quality Control for the Food Industry. 3rd ed. Vol.1. AVI Publishing Co., Westport, CT.

Pinder, A.C. and Godfrey, G. (Editors). 1993. Food Process Monitoring Systems. Chapman & Hall, London, New York.

Stauffer, J.E. 1988. Quality Assurance of Food: Ingredients, Processing and Distribution. Food & Nutrition Press, Westport, CT.

Thankamma, J., Food Adulteration, S.G, Wasani, New Delhi, 1976 The prevention of Food Adulteration Act, National Industrial Area, New Delhi, 1981.

Zeuthen, P., et al. (Editor). 1990. Processing and Quality of Foods. Chapman & Hall, London, New York.

Chapter - 10

Food Law and Regulation

Purpose of Law

The Food Laws or Regulations are made in order to protect people consuming these foods from undue risks, which may arise from processing, transport, retailing and consumption of food products which may undergo contamination, spoilage, inclusion of harmful chemicals/microbes or at harmful levels. Besides safety, consumers are also protected by these laws with respect to quality, quantity and substance that the food products are supposed to represent. Finally, laws also aim to protect the consumers from Nutrition and Health considerations with special consideration being given to vulnerable group such as infants, pregnant women etc.

Food additives are chemical substances added to foods to improve flavour, texture, colour, appearance and consistency or as preservatives during manufacturing or processing. Herbs, spices, hops, salt, yeast, water, air and protein hydrolysates are excluded from this definition.

In many countries, the use of food additives is regulated, and food additives must be declared on food labels by using their chemical names or numbers. the use of any new additive in a particular food, they ensure that:

General Principles of Food Safety Risk Management

1. Risk management should follow a structured approach
2. Protection of human health should be the primary consideration in risk management decisions
3. Risk management decisions and practices should be transparent
4. Determination of risk assessment policy should be included as a specific component of risk management
5. Risk management should ensure the scientific integrity of the risk assessment

process by maintaining the functional separation on risk management and risk assessment

6. Risk management decisions should include clear, interactive communication with customers and other interested parties in all aspects of the process
7. Risk management should be a continuing process that takes into account all newly generated data in the evaluation and review of risk management decisions

Challenges Food Regulation Law in Indian Food Industry

There are many challenges and opportunities for Indian food industry. Globalization has made available large number of choices of various products like fruit juices, beverages, biscuits, confectionery products etc. Many of the products are imported and some more are waiting. This scenario has both positive and negative aspects. People are ready for many new products so there is scope for industry. However, the imports will make domestic products to compete with them and at times superior imported product might dominate even with higher price. This presents a threat perception for Indian industry.

There are possibilities of exports for Indian products, but at the same time there is need for compliance with such WTO agreements like SPS and TBT as well as Codex standards. These impose additional quality parameters to the industry geared to Indian standards.

Science and technology is making advances at rapid rate especially in the fields of Biotechnology and Genetically Modified Foods are making appearance abroad as a result. These have superior properties both from agricultural points such as disease resistance, minimal susceptibility to damage during shipping, etc. as well as consumer view points like better flavour, colour, etc. They also may cost less to produce as the losses are less. There are other technologies like irradiation and packaging innovations also contribute to better quality and lesser losses. Those who take advantage of these will have greater advantage in the market. GM Foods are not allowed in many countries and these and their by-products may require authentication, which might be a negative point.

Consumers are also forcefully expressing their preferences and concerns about food products with respect to quality, safety, nutrition etc. These concerns also may make an impact on the market scenario. Since the markets have become connected due to globalisation, speed in marketing a new product is critical as competitors may quickly try to emulate a successful product.

Limitations of Current Food Laws

While the market scenario is not very friendly but highly competitive although there are many opportunities, any lack of support due to limitation in food laws is bound to make a negative impact on the industry. At present, there are many laws for same food products are at times they overlap or contradict one another. Many of the laws and standards are quite rigid and inflexible. There are rules, which lay down standards for hundreds of products with very little scope for innovation. This denies consumers the choice, which then will be provided by imported products and the market will be lost for Indian industry.

There is also weak enforcement and at times the implementation is ad-hoc and not uniform. This creates uncertainty among the industry about compliance, which may result in unjust harassment and threat of prosecution. This is not a healthy state for good growth of industry.

Integrated Food Law

Integrated Food Law is imperative for just and focused enforcement as well as for healthy growth of industry. Many countries have unified food laws including USA, Malaysia, Thailand, Indonesia and Pakistan. There are examples of groups of countries that have come together and formulated unified common food laws. Examples are Australia and New Zealand as well as European Union countries. They not only have integrated laws but also laws conducive for healthy industry producing safe and high quality products. In many of these there is focus on in-process quality control rather than product testing. Enforcement is more interested in food analysis lab.

National Food Law

- The Prevention of Food Adulteration Act, 1954
- The Fruit Products Order, 1955
- The Milk and Milk Products Order, 1992
- The Meat & Meat Products Order, 1973
- The Vegetable Oil Products (Control) Order, 1947

Legislations and Institutional Set-up

- Ministry of Agriculture
  - Insecticide Act
  - AGMARK
- Ministry of Health & Family Welfare
  - Prevention of Food Adulteration Act 1954
- Ministry of Food Processing Industries
  - Fruits & Vegetable Products (Control) Order
  - FPO 1955

- Ministry of Commerce
    - Export (Quality Control & Inspections) Act 1963
- Ministry of Civil Supplies, Consumer Affairs and Public Distribution
    - Standards of Weights & Measures Act
    - Standards of Weights & Measures (Enforcement) Act
    - Solvent Extracted Oils, De-oiled Meal and Edible Flour Control Order, 1967
    - Vegetable Products Control Order, 1976
    - Bureau of Indian Standards (BIS) Act 1986

Minimum requirements for Fruit Product Order (1956)

- Sanitary and hygienic conditions of premises, surroundings and personnel.
- Water to be used for processing.
- Machinery and equipment.
- Product standards.
- Maximum limits of preservatives, additives and contaminants have also been specified for various products

Central Fruit Products Advisory Committee

- Officials of concerned Government Departments
- Technical experts
- representatives of CFTRI
- Bureau of Indian Standards
- Fruit and vegetable Producers
- Processing Industry

ISO (International Organization for Standardization)

- International Non-Governmental organisation
- Federation of 123 National Standards Bodies
- Development, Approval and Promulgation of consensus based international standards
- 200 technical committees— about 650 sub-committees, 2000 working groups
- ISO 9000 to ISO 20000 series standards

The New Food Bill: August 2006

The scope of the Act has been widened a little to take care of some of the lacunae in the PFA Act. In the new Act, packaged water has been included as previously only the bottled mineral water was included in the PFA. Secondly, foods for special dietary uses or functional foods or nutraceuticals or health supplements etc. have been included in the Act. Earlier there was confusion since the enforcing agencies included these under drugs and wanted the drug license for the manufacture and marketing of these products.

This new Act should ensure the orderly development of food industry, as well as consumer safety by setting appropriate standards and uniform enforcement at the state level. It specifies offences and penalties for these offences, which are quite distinct and specific when compared with PFA. In order the proper regulation of the Act a separate body called Food Safety & Standards Authority of India (FSSAI) will be created. This body will have a full-time chairperson and will have representatives from Central and state governments, scientific bodies like CSIR & ICAR, Food & Beverage industry, consumer organizations, with a third of it to be represented by women. Its functions include, considering and approving act, rules and regulations, lay down standards, evolve guidelines for enforcement and supervision of implementation.

Among other features of the Act, Central Advisory Committee will be created to ensure close cooperation between the Food Authority and the enforcement agencies. It will have a CEO, who will be the legal representative in the Food Authority (FSSAI). There will be several scientific panels created of the subject matter experts for products and for consultations. They will also accredit laboratories. Scientific committees will be created which will provide opinions on multi-sectoral issues falling within the competence of more than one scientific panel and on issues not falling within the competence of any scientific panel.

Besides the general provisions on food safety, the Act provides for guidelines on analysis of food and offences and penalties under the Act. There is also provision for adjudication and Food Safety Appellate Tribunal to look into disputes arising out of the above.

Food Safety

Food safety in India is ensured by Government of India's Ministry of Health under the provisions of Prevention of Food Adulteration Act & Rules. They are responsible for Food Laws and the rules therein. State government Food & Drug Administration (FDA), which carries out surveillance using food inspectors, does the enforcement. There are food analysis labs, both state and central, which verify the authenticity of food products.

Any food safety legislation or standard requires involvement of several aspects including Research & Development, Information & Documentation, Education &

Training, Quality Assurance Program, Codex & International Norms, Advisory System, Planning, Enforcement and Surveillance. Various activities take place at different places such as education & research institutions, government laboratories, data bases including international & national, industry production and quality evaluation centers, and finally state level enforcement and surveillance departments. Due to the complex nature, any change is standards and enforcement has to be properly planned and executed after careful consideration of all these factors.

Food Additives: Regulations

There are different sets of regulations everywhere. Each country has its own set of rules for regulating food additives for example; US FDA Guidelines & Regulations gives the American regulations for food additives. Thus anyone producing and marketing food products in the US must abide by them. India has its own set of regulations under Prevention of Food Adulteration (PFA) Act & Rules. Each country has a set of regulations. When an Indian company wants to export to US, then it will have to follow the US regulations. When it wants to export to Australia their rules have to be followed. So there might be difficulties trying to follow many sets of regulations.

A group of countries may have a common regulation for example, European Union Directives, which give regulations for countries affiliated to it. This allows free exchange of food products across those EU countries. It avoids confusion because of many different regulations being followed for different countries. For international trade we have Codex, SPS, TBT regulations.

Under the WTO agreements, common regulations have been arrived at for those countries signatories to the agreement and this allows the international trade without much problems. FAO/WHO have come up with Codex rules, which are accepted by these countries.

Food Safety Administration [FSA]

FSA will be created for evolving guidelines for implementation and uniform enforcement of the Act and Rules and will be headed by an administrator not below the rank of a Joint Secretary to the Government of India. FSA will create training modules for central and state level food law enforcement officers. It will also ensure standardisation of operating procedures and development of infrastructure of accredited labs and analytical systems.

National Food Security to Global Food market

After the opening of the borders at the conclusion of WTO talks, the race for Global Market share is hitting up. Indian markets hitherto not very easily accessible for international players are considering opportunity in India. Unless we move rapidly forward, we are not only going to miss opportunities now available in international trade but we may have problems in domestic markets too as there will be international players coming into the competition. Currently, our processing, especially of value added processing, is abysmally low with less than 2% and there are very high wastages.

India has the potential to the world's largest producer of processed foods due to its vast resources in terms of agricultural produce. The existing food laws and regulatory framework impedes the growth of Indian food industry and stifles innovation. Regulatory mindset needs to shift from the current prescriptive vertical standards & regulations to horizontal norms and guidelines. We have the opportunity now to change for better or we may miss the bus.

Risk Assessment Policy

In these policies, animal models are relied upon with certain assumptions to establish potential human effects. A 100-fold safety factor is used for many assumptions and variations between species. The policy does not assign any ADI to additives, drugs and pesticides that are found to be genotoxic carcinogens. This permits some of these contaminants to be at levels "as low as reasonably achievable" (ALARA).

It must be remembered that all substances (chemicals) are poisons. There is none that may not act as poison; only the right dose differentiates a poison and a remedy. Some of the known toxins are at times given as remedy and some of the nutrients and medicines at very high levels are toxic. No food substance is unequivocally safe or unsafe. The safety depends both on the amount in the diet and on level of its exposure. It is also important to know that both natural and synthetic additives must be considered from safety aspects.

If no adverse health effects are demonstrated on the requested use, PFA will approve the food additive and recommend a **maximum level of food additive** permitted in particular foods. Generally, food additives approved by FSSI are safe to consume without any adverse reactions, however, some people are sensitive to particular food additives in common use.

Tips for Diagnosis and Management

1. Keeping a list of suspected foods allows easier identification of the culprit foods.
2. Record all foods consumed, and any symptoms that occur coincident with ingestion, in a food diary.
3. Showing the food diary to physicians will aid the diagnosis of adverse reactions to certain foods.
4. To confirm allergic reactions to certain foods, a skin prick test (SPT) or radio allergo sorbent test (RAST) should be performed.
5. If the test confirms an allergic reaction, avoid the substance that caused it.
6. Read the food label carefully and beware of the allergenic agents in food products.

References

Concon, J.M.1988. Food Toxicology. Marcel Dekker, New York

Desrosier, N.W. and Desrosier, J.N. 1977. Technology of Food Preservation.4th ed. AVI Publishing Co., Westport, Conn.

Farrer, K.T.H. 1987. A Guide to Food Additives and Contaminants. Parthenon Publishing Group, Park Ridge, NJ.

Food and Agriculture Organization of the United Nations and World Health Organization. 1992

Food and Drug Administration. 1990. Food Risk: Perception vs. Reality. Food and Drug Administration, U.S. Public Health Service, Dept. of Health and Human Services, Rockville, MD.

Food Protection Committee. 1981. Food Chemicals Codex. 3rd ed. National Academy of Science- National Research Council, Washington, DC.

Francis, F.J. 1992. Food Safety: The Interpretation of Risk. Council for Agricultural Science and Technology, Ames, IA.

Janssen, W.F. 1985. The U.S. Food and Drug Law: How it came, How it Work. U.S.Dept. of health and Human Services, Food and Drug Administration, Rockville, MD.

Schultz, H.W. 1981. Food Law Handbook. Chapman & Hall, London New York.

The Prevention of Food Adulteration Act, Naraina Industrial Area, New Delhi, 1981.

Thonney, P.F. and Bisogni, C.A. 1992. Government regulation of food safety: Interaction of scientific and societal forces, Food Technol. 46(1), 73-80.

Watson, D.H. 1993.Safety of Chemicals in Food: Chemical Contaminants. Ellis Horwood, New York.

Chapter - 11

Project Preparation

Introduction of Food Industry

India is one of the major food producers in the world. The food sector contributes to about 28% of India's GDP. India stands at 1st position in the world for production of cereals, milk, livestock, banana and Mango, 2nd in producing fruits and vegetables and ranks amongst top 5 in producing rice, wheat, groundnut, tea, coffee, tobacco, spices, sugar and oilseeds. India's share in global production of fruits is 10% and vegetables are 13.7%. The current consumption of fruits and vegetables is approx Rs.2 lakh crores at current prices with an estimated growth rate of 11% per annum. The growth rate is higher than cereals and milk and comparable to meet consumption

Status of Food Processing Industry

The food processing industry in India is still in a sorry state. The rural population comprising 70% including small cities, consume less than 10% of the processed foods and vegetables, whereas 60% of the processed food is consumed in four major metropolitan cities and 30% in the state capitals and big cities. Another fact is that 40% of the processed food and vegetables produced in the country in terms of value are bought by institutional buyers like Hotels, Restaurants and Defense etc. The highest growth in domestic market has been in fruit drinks, tomato ketch up and Jams.

There is another fact that India is the largest milk producer in the world, however, organized industry accounts for less than 15% of the milk produce in India. It is estimated that there may be a total production of 1100 million tons of production of food products mainly food grains, oilseeds, sugarcane and fruits/vegetables during 2011-12 and leaving marketable surplus of 870 million tons.

The demand for high value commodities particularly fruits; vegetables and milk would go up significantly during 2010 and 2020 in India. It is expected that the demand for fruits would go up from 56 million tons to 77 million tons (2010-2020),vegetables 113 to 150 million tons (2010-2020) and 104 to 143 million tons (2010- 2020) for milk, as projected by IARI.

Table 42: Projections of Marketable surplus

Commodity	Production*		Marketable surplus*	
	2001-02	2011-12	2001-02	2011-12
Food grains	213	321	110	166
Oilseeds	21	46	16	37
Sugarcane	297	433	276	402
Fruits and vegetables	133	300	166	265
Total	**664**	**1100**	**518**	**870**

Source: IARI *million of ton

The processed fruits and vegetables in India has been growing at about 9% per annum with the highest growth being witnessed by juices and ready to eat vegetables

Table 43: Status of Processed fruits and vegetables Industry in India

Category	Industry size (Rs in crores)		Key players
	Organized	Unorganized	
Jams	90	50	HLL, Mapro, Marico,Malas
Pickles	150	1000	Priya food, Praveen, Desai brothers, CavinKare, GD Foods
Sauce/ketchup	100	400	HLL, Nestle, GD foods, Heinz
Pulp/concentrates	100	-	Foods & Inns, BEC, clean foods, JainIrrigation, Usha International
Juices/fruit based drinks	500	-	Pepsi, Dabur, Godraj, Mother Dairy
Squashes	130	250	Kissan, Haldiram, Mapro
Ready to eat vegetables	100	-	Tasty bite, ITC, MTR
Potato chips	250	300	Pepsi, Haldiram, ITC
Cooking pastes	30	-	Dabur, HLL

Source: Rsereve Bank of India

Status of World Production Vis-à-vis Consumption of Fruits and Vegetables

The global production of fruits and vegetables is approx 1.7 billion tons and has grown at a CAGR of 3.4%. The China dominates the production of fruits and vegetables with 1/3rd of total global production. China, India, USA and Turkey are jointly responsible for 2/3rd of global vegetable production. The production of fruits and vegetables worldwide is hugely fragmented. Majority of food and vegetables is considered fresh. In low income markets, fruit consumption is mostly fresh, only 10% is consumed processed, whereas, in High income markets, 50% of fruit consumption is in the processed form. Besides consumer demand, triggers for food and vegetables processing are from the foodservice industry.

Fresh cut produce is a relatively new phenomenon and is a premium priced. Globally, fruits and vegetables are consumed close to the place of production. The global fruits and vegetables trade accounts for 5% of the global production and is currently approx. between 80-85 mn tons. However, the trade in fruits and vegetables is growing rapidly than trade in any other agriculture commodity. The fruits accounts for 60% of the total F&V trade and Banana is the world's most traded fruit. The banana, citrus, apples and pears account for 70% of global fruit trade. The vegetables trade is more regional, because of limited shelf life, but China is a dominant player in vegetable trade due to cost advantages and proximity to key import markets. The major key traders in the world for food and vegetables are EU and Mexico (Exporter) whereas USA and Japan are major importer.

What is Agri Business

Dealing in any agriculture output viz cereals, pulses, oilseeds, Horticulture, Floriculture, spices, plantation crops, livestock, poultry, marine products etc. is considered as Agri. business. Besides this, Beekeeping (Honey production), Dairying & Milk products, oil extraction, flour, derivatives, spices/Tea/Coffee Meat/Seafood/ Poultry processing is also covered under Agri. business. Since, Agri-business has a wider perspective, this paper limits to fruits and vegetable (Horticulture) related business opportunities.

The Agri. business pertaining to fruits and vegetables has three kinds of business components:

- Procurement/sorting /grading-transport (sell of fresh produce)
- Development of cooling facilities in terms of warehousing/storage/ transportation /IQF (Individual quick freezing)- Infrastructure support
- Processing of fruits and vegetables

Structure of Agri- Business

The structure of Agri-business has been precisely depicted in the following diagramme. This involves procurement/Aggregation of farm output, sorting, grading, logistics to processing units or to domestic markets, retail institution sale, export etc. Here, the very purpose is to eliminate the intermediaries to reduce the time, money and deterioration of quality of the produce.

Govt. Support for Horticulture business

The Horticulture Board through Govt. of India provides technical, financial, logistics, MIS, consultancy and Marketing support. A brief of the support offered on Horticulture business by Govt. of India

I. Scope of the Projects

Fruits and Vegetables Processing, Dehydration, Canning & Preservation Based Projects

India ranks first in the world in production of fruits and second in vegetables, accounting roughly 10 and 15 per cent, respectively, of total global production. India have a strong and dynamic food processing sector playing a vital role in diversifying the agricultural sector, improving value addition opportunities and creating surplus food for agro-food products. Presently, a mere 2.2 per cent of fruits and vegetables are processed, even as the country ranks second in the world in terms of production. This is comparatively low when compared to other countries like Brazil (30 per cent), USA (70 per cent) and Malaysia (82 per cent).The National policy aims to increase the percentage of food being processed in the country to 10 per cent by 2010 and 25 per cent by 2025. Food processing adds value, enhances shelf life of the perishable agro-food products and encourages crop diversification.

Major vegetables grown are Potato, Onion, Tomato, Cauliflower, Cabbage, Bean, Cucumber, Gherkin, Peas, Garlic and okra. The major fruits grown in India are Mangos, Grapes, Apple, Apricots, Orange, Banana Fresh, Avocados, Guava, Lichi, Papaya, Sapota and Water Melons. Mango, accounts for 40 percent of the national fruit production and India is one of the leading exporters of fresh table grapes to the global market. The changing food habits are discernible. There has been a positive growth in ready –to-serve beverages, fruit juices and pulps, processed fruits and vegetables products, i.e., dried or preserved and dehydrated vegetables and fruits such as sauces, preserved onions, cucumbers and gherkins, green pepper in brine, dehydrated garlic and ginger powder, dried garlic and ginger, tomato products, pickle and chutneys, processed mushrooms and truffles and curried vegetables.

The market size of the food processing industry is likely to increase from Rs 4600 billion in 2003-04 to Rs 8200 billion in 2009-10, and to Rs 13,500 billion in 2014-15.In the coming years India's share in the global processed food industry will get a raise from one per cent to three per cent. Indian food processing activity is still largely based on primary processing, which accounts for 80 per cent of the value addition. The government of India has set up a separate full-fledged ministry named "Ministry of Food Processing Industries "for the development and promotion of food processing industries. To boost fruits and vegetables processing, the ministry is extending financial support for setting up new units, modernization and upgradation of existing units. The Indian Food processing industry will continue to prosper, thanks to the rising income levels and modernized food retail stores. The food processing sector is likely to be the driving seat for the Indian economy.

We can provide you detailed project reports on the following topics. Please select the projects of your interests. Each detailed project reports cover all the aspects of business, from analysing the market, confirming availability of various necessities such as plant & machinery, raw materials to forecasting the financial requirements.

The scope of the report includes assessing market potential, negotiating with collaborators, investment decision making, corporate diversification planning etc. in a very planned manner by formulating detailed manufacturing techniques and forecasting financial aspects by estimating the cost of raw material, formulating the cash flow statement, projecting the balance sheet etc.

We also offer self-contained Pre-Investment and Pre-Feasibility Studies, Market Surveys and Studies, Preparation of Techno-Economic Feasibility Reports, Identification and Selection of Plant and Machinery, Manufacturing Process and or Equipment required, General Guidance, Technical and Commercial Counseling for setting up new industrial projects on the following topics.

Many of the engineers, project consultant & industrial consultancy firms in India and worldwide use our project reports as one of the input in doing their analysis.

A. Project on Manufacturing of Breakfast Cereal Foods

Project cost : Rs. 1650 lakh

1. Product and its Applications

Cereal processing is one of the oldest and most important food processing technologies and forms a part of the food production chain. Today, the cereal processing industry is as diverse as its range of products. Practically every meal contains cereal in one form or the other.

Extrusion cooking is a novel method for manufacturing food products from snacks and breakfast cereals to baby foods. However, as a complex multivariate process it requires careful control. Extruded foods are not just for kid snacks anymore. Today's extruders have opened the door to a global array of new products. The food makers around the world are discovering new ways of shaping, forming, squeezing and puffing foods to create healthful substance. The extruders are producing crispy flat bread, baby food, pet food, high fiber products, puffed snacks, pellet snacks, ready-to-eat breakfast cereals in puffed and flaked forms, meat analogues, pasta, cheese snacks and instant drinks and soups.

Extruded breakfast cereal products such as corn flakes, rice flakes, wheat flakes and other formulated breakfast cereal product are quite popular in India. Corn, rice, wheat are the major raw materials required for the production of extruded breakfast cereals. These ready – to - eat breakfast products are widely consumed in urban areas particularly by office going persons and school children. An extruder works by taking a blend of raw ingredients in one end and subjecting it to high heat and pressure in a cylinder. As the mix passes through the extruder cooker, it is shaped and fully or partially cooked. It takes its final shape as it is forced through a die at the exit end.

2. Market Potential

As these extruded foods are tasty and nutritious, the consumption is increasing day by day. There is also a good export potential for these products. The product is a household breakfast item which is in great demand in retail outlets though provision stores, bulk outlets to hotels, school- or office / industrial canteens, exports to under developed countries. A large number of new manufacturing units have been established recently catering to the need of consumers.

3. Basis and Presumption

1. The unit will work for 300 days per annum on single shift basis.
2. The unit can achieve its full capacity utilization during the 3rd year of operation.
3. Wages for skilled workers are taken as per prevailing rates in this type of industry.
4. Interest rate for total capital investment is calculated @ 12% per annum.
5. The entrepreneur is expected to raise 20-25% of the capital as margin money.
6. The unit would construct its own building.
7. Costs of machinery are based on average prices of machinery manufacturers.

4. Implementation Schedule

Project implementation will take a period of 12 months. Break-up of the activities and relative time for each activity is shown below:

Scheme preparation and approval	01 month
MSI provisional registration	3-4 months
Sanction of financial supports etc	4-6 months
Installation of machinery and power connection	6-12 months
Trial run and production	01 month

5. Technical Aspects

A. Location

The plant can be located at any suitable place keeping in view the marketing convenience, availability of power, water and skilled manpower.

B. Salient Features of Process / Technology

Breakfast cereals are produced by extrusion cooking method. For this purpose single screw and twin screw extruders are available. Twin screw extrusion cooking method is used for producing high quality products with better texture and precise process control. Ingredients are mixed and conveyed to a twin screw cooker extruder which gelatinizes the starch in cereal. The cooking temperature varies from 120° to 180° C for 10-15 minutes. Flavour, colour and vitamins are added at this stage. The cooked cereal is passed into a forming extruder wherein the mass is cooled and formed into pallets. The pallets are conditioned before flaking/ shredding. The flakes are then roasted and required coating of sugar is applied along with vitamins and minerals. This product is packed and dispatched.

Product Quality Specifications: Cornflakes are prepared from dehulled, degermed and cooked corn by flaking, partially drying and roasting in the form of crisp flakes of reasonably uniform size and golden brown in colour. It should be free from dirt, insects, larvae and impurities, any other extraneous matter.

6. Pollution Control

There is no major pollution problem associated with this industry except for disposal of waste which should be managed appropriately. The entrepreneurs are advised to take "No Objection Certificate" from the State Pollution Control Board.

7. Energy Conservation

Proper care should be taken while utilising the fuel for. There should be no leakage.

8. Production Capacity

Quantity	840 tpa
Installed capacity	1200 tpa
Optimum capacity utilization	70%
Working days	300
Manpower	29
Motive Power	90 kWh

9. Financial Aspects

I. Fixed Capital

A. Land & Building Amount (Rs. lakh)

Particulars	**Amount**
Land 4,500 m 2 & land development	098
Built up area 3,000 m 2	425
Total cost of land and building	**523**

B. Machinery and Equipment Amount (Rs. lakh)

Description	Amount
Plant, machinery consisting of Twin Screw Cooker Extruder, Forming Extruder, Pre conditioner, S.S. Mixer, Conveyer Belt Dryer, Storage Bins, Flavour Applicator, Packing Machinery, miscellaneous assets, etc.	780
Erection & electrification of machinery & equipment @10% cost	078
Office furniture & fixtures	032
Total	**890**

C. Pre-operative Expenses Amount (Rs. lakh)

Consultancy fee, project report, deposits with electricity department etc	**057**

D. Total Fixed Capital Amount (Rs. lakh)

(a+b+c)	**1470**

II. Recurring Expenses Per Annum

A. Personnel Amount (Rs. lakh)

Designation	No.	Salary	Amount
Factory Manager	1	30000	03.60
Managers	**2**	18000	04.32
Supervisory staff	**3**	15000	05.40
Office Assistant	**5**	13000	07.80
Technician	**3**	12000	04.32
Skilled workers	7	5000	04.20
Unskilled workers	18	3500	07.56
			37.20
Perquisites @15 %			05.80
Total	29		43.00

B. Raw Material including packaging materials Amount (Rs. lakh)

Particulars	Qty (t)	Rate/t (Rs.)	Amount
Raw Material including packaging materials	LS	LS	587.00
Total			**587.00**

C. Utilities Amount (Rs. lakh)

Particulars	Amount
Power	21.70
Water	01.30
Total	**23.00**

D. Other Contingent Expenses Amount (Rs. lakh)

Particulars	Amount
Repairs and maintenance of M&E @ 3 %	25.74
Repairs and maintenance of building @ 1%	04.25
Repairs and maintenance of site development @1%	00.98
Consumables & spares, others	16.50
Transport & travel	02.50
Publicity, postage, telephone	09.00
Insurance @1%	12.03
Total	**71.00**

E. Total Recurring Expenditure Amount (Rs. lakh)

(a + b + c + d)	**724.00**

III. Working Capital Amount (Rs. lakh)

Recurring expenses for 3 months	**180.00**

IV. Total Capital Investment Amount (Rs. lakh)

Fixed capital (Refer I e)	1470.00
Working capital (Refer III)	0180.00
Total	**1650.00**

V. Financial Analysis

A. Cost of Production (Per Annum) Amount (Rs. lakh)

Recurring expenses (Refer 9.2.5)	724.00
Depreciation on building @ 3.33%	014.15
Depreciation on machinery @10%	078.65
Depreciation on furniture @ 20%	003.20
Interest on Capital Investment @ 12%	198.00
Total	**1018.00**

B. Sale Proceeds (Turnover) per year Amount (Rs. lakh)

Item	Qty (t)	Rate/t (Rs.)	Amount
Breakfast cereals	840	1.80	1512.00

(ii) Net Profit (per year)

B.E.P. = Sales - Cost of production

$$= 1512 - 1018$$

$$= \text{Rs. } 494 \text{ lakh}$$

C. Net Profit Ratio

$$\text{B.E.P.} = \frac{\text{Net profit} \times 100}{\text{Sales}}$$

$$= \frac{494 \times 100}{1512}$$

$$= 32.6\ \%$$

D. Rate of Return on Investment

$$\text{B.E.P.} = \frac{\text{Net profit} \times 100}{\text{Capital Investment}}$$

$$= \frac{494 \times 100}{1650}$$

$$= 30\%$$

E. Annual Fixed Cost Amount (Rs. lakh)

All depreciations	096.00
Interest	198.00
40% of salary, wages, utility, contingency	054.80
Insurance	012.03
Total	**360.83**

F. Break Even Point

$$\text{B.E.P.} = \frac{\text{Annual Fixed cost} \times 100}{\text{Annual Fixed Cost} + \text{Profit}}$$

$$= \frac{360.83 \times 100}{360.83 + 494}$$

$$= 42\%$$

B. Project on Manufacturing of Protein Rich Biscuits

1. Product and Applications

Lifestyle is changing in most of the urban cities very rapidly. With both the husband and wife working the eating and cooking habits are undergoing changeover. Couples do not find time to prepare snacks at home and generally prefer snack food available in the market under different brand names. Bakery products have been popular and are being made since long. Bread and Biscuits are one such item equally preferred by children and grown up. The biscuits generally are prepared from wheat flour with various additions for flavor and taste. An addition of Soya flour increases its protein content and serves as a source of energy and nutrition. This value addition to the wheat flour does not affect its shelf life and economics. The technology for such biscuits is available with CFTRI. Compliance with PFA Act for such a unit is essential.

2. Industry Profile and Market Assessment

Bakery is an age old industry. Bread and biscuits in different varieties manufactured either large multinationals or at village level in small setup form the bakery products. The products are popular both at rural as well as urban level only the product and price differs. While the rural population prefers the cheap homemade variety the urban elite go in for costly varieties in different taste and assortments. There is market for both the varieties. The biscuits in general sense mean a product with lot of calories and which is generally consumed as a snack at tea time or children consume it in between meals. There are variations in the taste and flavour of biscuits as it may be chocolate flavour or orange, vanilla, straw berry etc it may be with cream or without cream. The biscuits though have calories but lack protein.

This profile seeks to introduce biscuits with protein for calorie conscious and for those who wish to retain energy. Biscuits are consumed by all irrespective of age or income groups. With addition of soya flour the consumer can get additional nutrient as soya is high in protein but low in fat and carbohydrates. Soya products like milk, nuggets, paneer, flour, biscuits etc. are becoming popular and the industry is growing at a good rate. It has a good demand in urban areas and metropolitan cities. Once the product establishes its Brand, export opportunities can also be explored. Middle East countries and other western countries are places where it has demand.

3. Manufacturing Process & Know How

The process of manufacturing is simple and standardized. Various ingredients like flour/maida, soyabean flour, starch, soda, salt, preservatives, sugar, ghee, etc are thoroughly mixed with the help of water and properly kneaded dough is set on biscuit moulds and then baked in an oven. The baked biscuits are cooled, weighed and packed for dispatch.

Know how is available with Central Government research Laboratories.

The machinery is all indigenously available. The production capacity envisaged is 80 tonnes per year 300 days and two shifts basis working.

4. Plant and Machinery

The main plant and machinery required comprise

- Roller cutting m/c size 48 inches with oven complete with reduction gear electrical and accessories. - 1 no.
- Automatic screw type flour sifter with motors - 1no.
- Grinder 30 kg. - 1nos
- Roller sheeter with motors - 1 no
- Double action horizontal mixing machine 250kg - 1 no.
- Cooling conveyor 18" with motors. - 1 no
- Oil spraying machine - 1 no
- Turn table with motors - 1 no
- Syrup machine with motors - 1 no.
- Biscuit grinder - 1 no.
- SS tanks, trays, crates, weighing scales
    - The total cost of machinery is estimated to be Rs.13.85 lakhs.
    - The unit will also require miscellaneous assets such as furniture, fixtures, storage facilities etc. the total cost of these is estimated to be Rs. 0.75 lakhs.
    - The total requirement of power shall be 70 HP, the unit will need 3000 lits of water daily.

5. Raw Material and Packing Material

The basic raw material for the unit is wheat flour and soya flour. Other items like starch, salt, sugar, baking soda, colours, flavours shall be needed in small quantities. Packing material like boxes, polythene sheets, box strappings etc shall also be required. On an average the raw material cost has been estimated to be Rs.31.15 lakhs at rated capacity. At 60% capacity in 1st year the cost works out to Rs 18.69 lakhs.

6. Land and Building

For smooth operation of the unit, it will require 500 sq. mts of open land and a built up area of 300 sq. mts. The total cost of land and building is estimated at Rs. 9.00 lakhs.

7. Manpower

For smooth functioning of the unit the requirement of man power is expected to be around 8 persons.

Sales person	self
Skilled Workers	2
Semi skilled workers	2
Helpers	2
Supervisor	1

The annual salary bill is estimated to be around Rs.2.58 lakhs.:

8. Sales Revenue: (100% capacity)

Selling price varies depending on the product mix quality. An average price of Rs 85,000/- per tonne has been taken the annual income at installed capacity of 80 tonnes is Rs 68.00 lakhs.

9. Cost of Project

Sales	Rs. lakhs
Land & Building	9.00
Plant & Machinery	13.85
Other assets	0.75
Contingencies	2.30
P & P expenses	1.00
Margin money	1.35
Total	28.25

10. Means of Finance

Promoters Contribution	8.50
Term Loan	19.75
Total	28.25

11. Profitability: (60%capacity)

	Rs. lakhs
Sales	40.80
Raw material	18.69
Salary	2.58
Utilities	0.90
Stores & Spares	0.30
Repairs & Maintenance	0.36
Selling expenses	9.18
Administrative expenses	0.60
Depreciation	2.94
Interest on T.L	2.10
Interest on W.C	0.37
Cost of production	38.02
Profit	2.78

12. Requirement of Working Capital

		Margin	W.C	Margin Money
Packing material	15 days	30%	0.80	0.25
Stock of finished goods	15 days	25%	1.00	0.25
Working expenses	1 month	100%	0.40	0.40
Sale on credit	15 days	25%	1.80	0.45
Margin money for W.C				1.35

13. Break Even point 49%

14. Machinery Suppliers:

1. M/S Baker & Co P. Ltd, Crawford Market Mumbai 400 008.
2. M/s Raylon Metal Works J.B.Nagar, Andheri(E) Mumbai.
3. M/S Baker Enterprises,Near Peeragarhi, New Delhi.110 041

C. Project on manufacturing of Aseptic Juice Concentrate

Production Capacity:	:	Rs. 17.70 lakh

1. Product and its Applications

The aseptic fruit concentrates are prepared after evaporation of water from fruit juices. These maintain quality, prolonged shelf life and optimize the transport and storage cost.

The plant facility offers a wider range of final juice products for distribution such as fruit juice concentrates either packed in bulk or for consumer use, blended juices, both concentrated and ready for consumption, along with single strength fruit juices packaged in a wide variety of commercial containers. There has been a remarkable growth in the demand for the juice concentrates due to the increasing popularity of new non-alcoholic and alcoholic fruit drink products, ice cream, yoghurt and baby food etc. The fruit & vegetable syrups or concentrates are used as flavors in these products. The soft drink market is also creating huge demand for the concentrates. The move away from alcoholic drinks and the relative inconvenience of hot drinks has resulted in a major shift to packaged soft drinks of all flavours. In addition, the confectionery industry has followed the suit and new products are now constituted by fruit and vegetable concentrates as part of their confectionery formulations and processes.

Consumers demand healthy options while at the same time have increased concerns about food safety risks, two trends that may often conflict with each other in juice concentrate processing. To address the demand for juice concentrate without synthetic or chemical preservatives, the manufacturers are exploring new preservation methods like aseptic processing.

2. Market Potential

The fruit juice industry has made good progress in India . According to trade sources, the total market for fruit drinks & nectars has reportedly shown a growth rate of 10 -15% per annum in the past. The Indian market for fruit juices has reported an annual growth of 25-30%. The new sector which has potential to be explored is combination of various products like fruit and milk combination, fruit-yogurt drinks that are more natural & nutritious drinks.

At international scenario, the consumption of soft and fruit drinks is estimated to be quite high. France has registered an annual growth of 20% during in past few years. Per capita consumption of fruit juice in USA is 46 L, Germany 39 L, Holland 24 L, UK19 L, Spain15 L, Japan15 L, France 12 L & Italy 8 L. Per capita consumption of soft drinks is highest in Norway at 122.7 L, followed by UK 113 L, Sweden 92.2 L, Holland 90 L, Germany , Ireland & Austria 85 L each and France 62 L. These figures indicate potentiality of exporting fruit juice concentrates to European markets. Brazil is the world leader in export of citrus juice concentrates and nectars.

3. Basis and Presumption

1. The unit will work for 300 days per annum for 20 hrs working / day .
2. The unit can achieve its full capacity utilization during the 3rd year of operation.

3. The wages for skilled workers are taken as per prevailing rates in this type of industry.
4. Interest rate for total capital investment is calculated @ 12% per annum.
5. The entrepreneur is expected to raise 20-25% of the capital as margin money.
6. The unit would construct its own building.
7. Costs of machinery are based on average prices of machinery manufacturers.

4. Implementation Schedule

Project implementation will take a period of 8 months. Break-up of the activities and relative time for each activity is shown below:

Scheme preparation and approval	01 month
SSI provisional registration	1-2 months
Sanction of financial supports etc	2-5 months
Installation of machinery and power connection	6-8 months
Trial run and production	01 month

5. Technical Aspects

5.1 Location

The plant can be located at any suitable place keeping in view the marketing convenience, availability of power, water and skilled manpower.

5.2 Salient Features of Process / Technology

The industry and technology of processing fruit into concentrates and blended juices includes well-known individual steps and equipment for accomplishing each of the stages or steps in the fruit processing chain of events. Ripe fruits are cut and passed through a pulper to get a smooth pulp. The pulp is pasteurized, cooled and evaporated and passed through an aroma recovery system. The aroma is collected and added back to the concentrate. Manufacture of fruit juices and concentrates by canning and vacuum concentration has been in practice for many years. Further, there are chances of deterioration of product packed in cans through leakage, rusting, etc. leading to bacterial contamination. Therefore, new technology of aseptic filling and packing is being adopted worldwide for avoiding these chances. For extraction of fruit juice and pulp, hot break system has been recommended.

Evaporative concentration, freeze concentration and reverse osmosis are commercially exploited. As freeze concentration and membrane concentration are very expensive technologies, evaporating concentration has been considered. In evaporating concentration double-effect falling film evaporator has more advantages over single effect falling film evaporator.

After obtaining juice, it is subjected to vacuum concentration, Vacuum concentrator operate on vacuum with steam requirement of 570 kg/hr and water requirement of 5000 L/hr. In evaporator, juice is subjected to heating at 100°C under vacuum. Vacuum concentrator is equipped with aroma recovery unit. The exit temperature for concentrate is15°C and is allowed to cool to 10°C. The concentrates are then stored in No.10 cans, 200 L carboys or 220-240 kg drums.

For manufacturing baby food, puree or paste of banana, carrot, guava etc. are mixed together in mixing & standardization tanks, where other ingredients are added along with the required amount of water and preservatives. The paste is then homogenized, pasteurized and filled into cans or bottles. Baby food thus prepared is highly nutritious. Such baby foods are rich source of vitamin C, vitamin A., minerals, iron and other valuable nutrients.

6. Pollution Control

There is no major pollution problem associated with this industry except for disposal of waste which should be managed appropriately. The entrepreneurs are advised to take "No Objection Certificate" from the State Pollution Control Board.

7. Energy Conservation

The fuel for the steam generation in the boiler is coal or LDO depending upon the type of boiler. Proper care should be taken while utilising the fuel for the steam production. There should be no leakage of steam in the pipe lines and adequate insulation should be provided.

8. Production Capacity

Quantity	3 000 t
Installed capacity	20 tpd
Optimum capacity utilization	70%
Working days	220
Manpower	45
Utilities	
Motive Power kWH	180

9. Financial Aspects

9.1 Fixed capital

9.1.1 Land & building amount (Rs. lakh)

Particulars	**Amount**
Land 15,000 m 2 & development	090.00
Built up area 8,500 m 2 -	535.00
Total cost of land and building	625.00

9.1.2 Machinery and Equipment Amount (Rs. lakh)

Description	**Amount**
Fruit washer, fruit mill, pulper, blancher, evaporator, centrifuge, pasteurization plant, chilling plant and aseptic processing unit, weighing scales, allied equipment, Transport	480.00
Erection & electrification of machinery & equipment @10% cost	048.00
Office furniture, fixtures & miscellaneous assets	042.00
Total	**540.00**

9.1.3 Pre-Operative Expenses Amount (Rs. lakh)

Consultancy fee, project report, deposits with electricity department etc	045.00

9.1.4 Total Fixed Capital Amount (Rs. lakh)

(9.1.1+9.1.2+9.1.3)	1240.00

9.2 Recurring Expenses Per Annum

9.2.1 Personnel Amount (Rs. lakh)

Designation	**No.**	**Salary Per month**	**Amount**
Factory General Manager	1	60,000	07.20
Managers	3	35,000	12.60
Supervisory staff	6	22,000	15.84
Accounts, Admn staff	5	18,000	10.80
Technicians	4	15,000	07.20
Skilled workers	8	12,000	11.52
Unskilled workers	18	4,000	08.64
			73.80
Perquisites @15 %			11.00
Total	45		84.80

9.2.2 Raw Material Including Packaging Materials Amount (Rs. lakh)

Particulars	**Amount**
Raw material, additives, packaging material, miscellaneous	1820.00
Total	1820.00

9.2.3 Utilities Amount (Rs. lakh)

Particulars	Amount
Power	46.00
Water	04.20
Total	50.20

9.2.4 Other Contingent Expenses Amount (Rs. lakh)

Particulars	Amount
Repairs and maintenance @10%	054.00
Consumables & spares, others	068.00
Transport & travel	019.50
Publicity, postage, telephone	009.00
Insurance @1%	010.50
Total	161.00

9.2.5 Total Recurring Expenditure Amount (Rs. lakh)

(9.2.1 + 9.2.2 + 9.2.3 + 9.2.4)	2116.00

9.3 Working Capital Amount (Rs. lakh)

Recurring expenses for 3 months	530.00

9.4 Total Capital Investment Amount (Rs. lakh)

Fixed capital (Refer 9.1.4)	1240.00
Working capital (Refer 9.3)	0530.00
Total	1770.00

10. Financial Analysis

10.1 Cost of Production (per annum) Amount (Rs. lakh)

Recurring expenses (Refer 9.2.5)	2116.00
Depreciation on building @ 5%	0027.00
Depreciation on machinery @10%	0052.80
Depreciation on furniture @ 20%	0008.40
Interest on Capital Investment @ 12%	0211.80
Total	2416.00

10.2 Sale Proceeds (Turnover) Per Year Amount (Rs. lakh)

Item	Qty (t)	Rate/ton (Rs.)	Amount
Concentrate	3000	1.20 lakh	3600.00

10.3 Net Profit Per Year

10.3 Net Profit per year

= Sales – Cost of production

= 3600 – 2416

= Rs.1184 lakh

10.4 Net Profit Ratio

$$= \frac{\text{Net profit} \times 100}{\text{Sales}}$$

$$= \frac{1184 \times 100}{3600}$$

10.5 Rate of Return on Investment

$$= \frac{\text{Net profit} \times 100}{\text{Capital Investment}}$$

$$= \frac{1184 \times 100}{1770}$$

= 66.9%

10.6 Annual Fixed Cost Amount (Rs. lakh)

All depreciations	088.20
Interest	211.80
40% of salary, wages, utility, contingency	119.20
Insurance	010.50
Total	**429.70**

10.7 Break even Point

$$= \frac{\text{Annual Fixed cost} \times 100}{\text{Annual Fixed Cost} + \text{Profit}}$$

$$= \frac{429.70 \times 100}{429.70 + 1184}$$

$$= \frac{42870}{1613.70}$$

$$= 26.6\ \%$$

11. Addresses of Machinery and Equipment Suppliers

SSP Pvt. Ltd 19-DLF Industrial Area, Phase-II, 13/4-Mathura Road
Faridabad - 121 003 (Haryana)

Alfa Laval (India) Limited 30 to 33 & 74 & 82, A/P Sarole Veer Road ,
Tal - Bhor, Dist - Sarole, Pune-412 205
Web page: www.alfalaval.com

Mather & Plant (India) Ltd Chinchwad,
Pune-411019

FMC Asia Pacific , Inc . 1/F, Krislon House Saki Vihar Road ,
Saki Naka Mumbai , 400 072

M/s Bajaj Maschinen (P) Ltd. 7/20, 7/27, Jailaxmi Industrial Estate,
Site IV Sahibabad Industrial Area
Ghaziabad -201301
Email : bajaj@del3.vsnl.net.in/ bajajmachines@vsnl.com
Web: www.bajajmachinen.com

M/s Narangs Corporation P -25, Connaught Place
New Delhi -110001
Email: narangs@appexmail.com

M/s Larsen & Toubro Ltd. L &T House, Ballard Estate,
Mumbai -400001
Web: www.larsontoubro.com

12. Other Special Features

The facilities can also be utilised to manufacture fruit juices, fruit pulps, concentrates, fruit pastes & purees as well as vegetables juices & pulps for fuller utilisation of capacity.

D. Project on Manufacturing of Amla Products

1. Product and its Applications

A bitter fruit Amla is better known for its medicinal values. It is used in various forms and is an important ingredient in Ayurvedic and herbal medicines. With a shift in thinking and awareness about wellness Amla in different forms is consumed by health conscious people. Amla being a fruit is available during season for 4-5 months. It is rich in vitamins and it is processed to improve its shelf life. With the popularity of Ayurveda both as a medicine and as a beauty aid the demand for various products with Amla has increased.

Compliance under PFA Act is compulsory.

2. Industry Profile and Market Assessment

Amla products are available in different forms, powder, syrups and oils, dried and treated slices, crushed and pasty form (as in Chawanprash). Amla is also consumed in raw form as a fruit it has medicinal values as it is considered good for skin, it is good for digestion and offers cure for diabetics in controlling sugar level. Somee of its products like chawan prash and oils have a fairly large market with well established companies like Dabur, Zandu, Charak Baidyanath etc. competing to capture larger share of the market.

The advertising by these companies have created the requisite consciousness about the product but the prices of these products are sometimes beyond the reach of the common man. A small unit thus catering to its local area and keeping the prices with reach as well as maintaining the purity and quality of the product will be able to survive. The price and the quality are two important factors. A small unit with limited manufacturing capacity will also be send a message that its products are homemade and cater to demand only. The unit can also seek franchise from the reputed manufacturers to cater to their requirement in the area.The area selection for marketing and its pricing and quality are of utmost importance to capture a market.

3. Manufacturing Process & Know How

Amla processing is a well established process and in fact it is readily processed in many household. Fresh fully grown amlas are cleaned and washed. They are then cooked in pressure vessels for making pasty products or are sliced and dried in sun to make chewable slices. Once the fruit has been boiled seeds are removed before drying. The fruit is also sometimes crushed to extract its juice. Spices and other ingredients are added. The products are properly packed and dispatched. Know how is available with Central Government research Laboratories. The machinery is all indigenously available. The production capacity envisaged is 3000 kgs per year keeping in view the limited market.

4. Plant and Machinery

The product does not require many items of machinery. Keeping in mind the size of market production capacity of 3000kgs per year with working of around eight months as Amla is not available during off season of about four months. The main plant and machinery required comprise

- Mixer Grinder - 2 nos.
- Gas Bhatti
- Fruit Crates.
- Stainless steel utensils/Plastic tubs.
- Pressure cookers - 2nos.
- Weighing scales
- Bag sealing machines - 2nos.

The total cost of machinery is estimated to be Rs.45000/-.The unit will also require miscellaneous assets such as furniture, fixtures, storage facilities etc. the total cost of these is estimated to be Rs. 20,000/-.

5. Raw material and Packing Material

The basic raw material for the unit is good quality Amla fruit. After removal of seeds the actual recovery is 60% to 65%. Considering a recovery of 60% Amla required at 100% capacity is 5000kg. Amla is available for 5 months and requirement of subsequent 3 months is stored. Other items like salt, sugar, cumin, brahmi etc. are all available locally. The finished product is packed in plastic bags and the syrup is packed in plastic bottles. A proper survey has to be conducted for location of the plant for easy availability of raw material. The price of raw material, spices and packing material at full capacity utilization is estimated to be Rs. 1.08 lakhs per year. At 60% capacity in 1st year the cost works out to Rs 0.65 lakhs.

6. Land and Building

For smooth operation of the unit, it will require a small built up area of 40-45 sq. mts. The same may be taken on rental basis.

7. Manpower

For smooth functioning of the unit the requirement of man power is expected to be around 3 persons.

Skilled Workers	1
Helpers The annual salary bill is estimated to be around	2 Rs. 0.48 lakhs

8. Sales Revenue: (100% capacity)

Many varieties of products can be made from Amla. But it is suggested that to identify the exact product mix the market has to be ascertained. Products like slices, salted/sugared, powder, sticks etc. can be sold @Rs.50/- per kg. wheras syrups and morabas @ Rs.80/- per kg. Other products like Chawanprash and Hair oils have not been considered as they are high value items and need extra marketing efforts. Assuming that the two products will be equally sold the sales income at 100% capacity will be as under:

Product	**Qty(Kgs)**	**Price/kg(Rs.)**	**Value(Rs. Lakhs)**
Slices,Goli,Sticks	1500	80	1.20
Syrups, Morabbas	1500	120	1.80
Total			**3.00**

9. Cost of Project:

Particulars	**Rs. lakhs**
Land & Building	on rent
Plant & Machinery	0.45
Other assets	0.20
Contingencies & pre-expenses	0.25
Total	0.90

10. Means of Finance

Promoters Contribution	0.30
Term Loan	0.60
Total	0.90

11. Profitability : (60%capacity)

Sales	1.80 (Rs. lakhs)
Raw material	0.70
Salary	0.32
Utilities	0.08
Repairs & Maintenance	0.03
Selling & Admn expenses	0.18
Depreciation	0.07
Interest on T.L	0.07
Interest on W.C	0.04
Cost of production	1.49
Profit	**0.31**

12. Requirement of Working Capital

It may be difficult for Bank to finance working capital needs as the stock of raw material will not be much during season and the majority of the customers will be scattered retailers with small quantities. Hence it is suggested to avail a personal loan. The amount envisaged is Rs.30000/-

13. Break Even Point: 57%

14. Machinery Suppliers

1. M/S Sujata Enterprises Laxmi Road Pune.
2. M/s Techno Equipment, Parekh Street, Girgaon Mumbai.
3. M/S Hildon Packaging Machine P. Ltd. 16, MIDC, Chakala Andheri, Mumbai

E. Project on manufacturing of Carbonated Fruit Beverages

Production Capacity	:	6 lakh bottles of 200 mL cap./annum

1. Product and its Applications

Carbonated soft drinks are known for their thirst quenching and refreshing properties. However, they lack nutritional content. On the other hand, fruit juices, packed in cans, bottles or pouches are also very popular in the country which contain the nourishing properties and goodness of the fruit. Carbonated fruit based beverage is a new concept which provides nutritional elements of the fruit along with natural pigments and flavour in addition to carbonation effects.

2. Market Potential

There is well established market for carbonated soft drinks. The fruit juice/pulp based carbonated beverage is expected to beget a rousing response in the market. Like other fruits amla, a rich source of Vitamin C produces an excellent relishing drink.

3. Basis and Presumptions

1. The unit proposes to work at least 300 days per annum on single shift basis.
2. The unit can achieve its full capacity utilization during the 2nd year of operation.
3. The wages for skilled workers is taken as per prevailing rates in this type of industry.
4. Interest rate for total capital investment is calculated @ 12% per annum.
5. The entrepreneur is expected to raise 20-25% of the capital as margin money.
6. The unit proposes to construct its own building as per FPO requirements.
7. Costs of machinery and equipment are based on average prices enquired from machinery manufacturers.

4. Implementation Schedule

Project implementation will take a period of 8 months. Break-up of the activities and relative time for each activity is shown below:

Scheme preparation and approval	:	01 month
SSI provisional registration	:	1-2 months
Sanction of financial supports etc	:	2-5 months
Installation of machinery and power connection	:	6-8 months
Trial run and production	:	01 month

5. Technical Aspects

5.1 Process of Manufacture

Fruits like amla, grape, jamun, lime, phalsa can be used. Fully ripe sound fruits are selected. These are washed and juice/pulp extracted. Further process involves preparation of fruit syrup base, carbonation by post mix method, heat processing, cooling and packaging in 200 mL bottles.

5.2 Quality Control and Standards: As per FPO Specifications.

6. Pollution Control

There is no major pollution problem associated with this industry except for disposal of waste which should be managed appropriately. The entrepreneurs are advised to take "No Objection Certificate" from the State Pollution Control Board.

7. Energy Conservation

The fuel for the steam generation in the boiler is coal or LDO depending upon the type of boiler. Proper care should be taken while utilising the fuel for the steam production. There should be no leakage of steam in the pipe lines and adequate insulation should be provided.

8. Production Capacity

Quantity	6 lakh bottles /annum
Installed capacity	1.7 lakh Litres
Optimum capacity utilization	70%
Working days	300/annum
Manpower	17
Utilities	
Motive Power	15 kWH
Water	25 kL/day

9. Financial Aspects

9.1 Fixed Capital

9.1.1 Land & building

Particulars	**Amount (Rs. lakh)**
Land 400 m 2	01.10
Built up area 250 m 2	09.90
Total cost of land and building	11.00

9.1.2 Machinery and Equipment

Description	**Amount (Rs. lakh)**
Washing tank, fruit mill, hydraulic juice extractor, steam jacketing kettles (3),boiler, carbonation unit, mixing tank, pasteurizer, crown corking machine, pH meter, weighing scales, labelling machine, bottle washing unit, refractometer, laboratory equipment.	: 18.00
Erection & electrification of machinery & equipment @10% cost	: 01.80
Office furniture & fixtures	: 01.20
Total	: **21.00**

9.1.3 Pre-operative Expenses

Consultancy fee, project report, deposits with electricity department etc	**01.00 (Rs. lakh)**

9.1.4 Total Fixed Capital

(9.1.1+9.1.2+9.1.3)	**33.00 (Rs. lakh)**

9.2 Recurring Expenses Per Annum

9.2.1 Personnel

Designation	**No.**	**Salary Per month**	**Amount (Rs. lakh)**
Factory Manager	1	12,000	01.44
Production supervisor	1	08,000	00.96
Office Assistants	2	06,000	01.44
Mechanic/ Technician	1	05,000	00.60
Skilled workers	4	04,000	01.92
Unskilled workers	8	03,500	03.36
			09.72
Perquisites @15 %			01.46
Total	17		**11.18**

9.2.2 Raw Material Including Packaging Materials

Particulars	**Qty (t)**	**Rate/t (Rs.)**	**Amount (Rs. lakh)**
Fruits	40	10,000	04.00
Sugar	14	22,000	03.08
Citric acid	0.25	160,000	00.40
Bottles	6.05 lakh	4/ each	24.20
Crown corks	6.05 lakh	—	02.88
Labels	6.00 lakh	—	03.00
Chemicals	LS	LS	01.10
Cartons (24 bottles cap.)	25,000	14 each	03.50
Carbon dioxide gas	—	—	00.84
Total			**43.00**

Contd.

9.2.3 Utilities

Particulars	Amount (Rs. lakh)
Power	02.90
Water	00.20
Coal	03.90
Total	**07.20**

9.2.4 Other Contingent Expenses

Particulars	Amount (Rs. lakh)
Repairs and maintenance @10%	01.98
Consumables & spares, others	00.58
Transport & travel	00.45
Publicity, postage, telephone	00.80
Insurance @1%	00.31
Total	**04.12**

9.2.5 Total Recurring Expenditure

(9.2.1 + 9.2.2 + 9.2.3 + 9.2.4)	**65.50 (Rs. lakh)**

9.3 Working Capital

Recurring expenses for 3 months	**16.40 (Rs. lakh)**

9.4 Total Capital Investment

	Amount (Rs. lakh)
Fixed capital (Refer 9.1.4)	33.00
Working capital (Refer 9.3)	16.40
Total	**49.40**

10. Financial Analysis

10.1 Cost of Production (per annum)

	Amount (Rs. lakh)
Recurring expenses (Refer 9.2.5)	65.50
Depreciation on building @ 5%	00.45
Depreciation on machinery @10%	01.88
Depreciation on furniture @ 20%	00.24
Interest on Capital Investment @ 12%	05.93
Total	**74.00**

10.2 Sale Proceeds (Turnover) Per Year

Item	Qty	Rate (Rs.)	Amount (Rs. lakh)
1. Carbonated beverage in 200 mL bottles	600,000	11.00	66.00
2. Refund of empty bottles	600,000	3.50	21.00
Total			**87.00**

10.3 Net Profit Per Year

10.4 Net Profit Ratio

10.5 Rate of Return on Investment

10.6 Annual Fixed Cost

	Amount (Rs. Lakh)
All depreciations	02.57
Interest	05.93
40% of salary, wages, utility, contingency	08.80
Insurance	00.31
Total	**17.61**

10.7 Break Even Point

11. Addresses of Machinery and Equipment Suppliers

1. B.Sen Barry & Co.
   65/11, New Rohtak Road
   New Delhi – 110 005
2. Macneill and Magor Ltd.
   4, Mangoe Lane
   Kolkata – 700 001
3. Bajaj Maschinen Pvt. Ltd.
   7/20-7/27 Jai Laxmi Industrial Estate, Site IV,
   Sahibabad Industrial Area - 201010
   Dist.Ghaziabad, UP
4. M/s Gee Gee Foods & Packaging Co. (P) Ltd.
   B -188/2, Savitri Nagar, Malviya Nagar
   New Delhi -110017
   Email: swantour@satyam.net.in

5. M/s D.K. Barry & Co. (P) Ltd.
   11/35, West Punjabi Bagh,
   New Delhi -110026
   Email: dheerajbahry@rediffmail.com
6. M/s Azad Engineering Company
   C-83, B.S. Road , Indl Area,
   Ghazibad -201009
   Email: azadeng@satyam.net.in/ azadenggco@hotmail.com
7. M/s M. Son Industries
   D -33, Sector 2
   Noida -201301
   Email: mson@nda.vsnl.net.in
8. M/s Sanjivan Industries
   A -13, Karimsheth Indl. Estate,
   Wagdevi Nagar, Near Valishali Nagar
   Dahisar (E), Mumbai -400068
   Email: sanjivanind@roltanet.com
9. M/s Biotech & Food Systems (P) Ltd.
   435 –P, Sector 14,
   Gurgaon -122001
   Email: dharamwal@yahoo.com

F. Project on Manufacturing of Dehydrated Vegetables

1. Product and Applications

Vegetables are a seasonal product and cannot be stored over a long period. Hence majority of the vegetables are not available during off season. To overcome the problem dehydration technique has been developed by which vegetables can be preserved for longer period and consumed whenever needed as fresh vegetable. Thus ensuring supply during off season. This value addition to the vegetables besides ensuring availability also improves its economics. In the present day life style when both the members of family work and do not find time to shop for fresh vegetables, clean them, sort them and size them before cooking, dehydrated vegetables are handy. Some of the popular vegetables such as peas, cauliflower, spinach, carrots etc. can be dehydrated and preserved for consumption throughout the year. The technology for dehydration is available with CFTRI. Compliance with PFA Act for such a unit is essential.

2. Industry Profile and Market Assessment

In the Indian households vegetables are cooked every day and in fact separately for lunch and dinner. This exercise is time consuming and laborious as fresh vegetables are to be procured cleaned, sorted and cooked. In the present day fast life the housewife can hardly spare time for such activity along with their career. Dehydrated vegetables are thus a solution to save time and ensure availability of vegetables for meals. Dehydration of these vegetables will ensure their availability throughout the year. Further cooking dehydrated vegetables is also not troublesome as the pre cooking steps of sorting, cutting, sizing are eliminated and the vegetable is just soaked in water to hydrate it and then cooked. It is easier to store and transport.

The major limitation with the bulk of Green vegetables is that they are grown in a limited period only lasting for 3-4 months and thus their availability is restricted to this period. Dehydration of vegetables results in its compactness and weight reduction thus it becomes easier to handle the product. This also helps in exporting the product to other countries where ever Indian cuisine is popular. It has a good demand in urban areas and metropolitan cities. Once the product establishes its Brand export opportunities can also be explored. Middle East countries and other western countries with Indian population are places where it has demand.

3. Manufacturing Process & Know How

The process of manufacturing is simple and for the purpose of this profile vegetables like cabbage, cauliflower, spinach, and carrots have been considered. In case of cauliflower the vegetable is chopped to make small pieces and washed. The pieces are then blanched and dried in cold air. Spinach leaves are separated from stalk, washed and dried in drier. Carrots are washed scrapped and cubed after washing. The cubes are blanched and dried. These dehydrated vegetables are packed and stored. Packing is very critical as any fungal growth would damage the product. Process and weight loss varies from product to product but on an average is 25% as the vegetables are dehydrated. Know how is available with Central Government research Laboratories. The machinery is all indigenously available. The production capacity envisaged is 400 tonnes per year in 2 shifts and 300 days working.

4. Plant and Machinery

The main plant and machinery required comprise

- Washing Tanks with sets of cubers & slicers - 2 nos.
- Blanching Tank with Thermostat - 1nos.
- Stacking Trays.
- Pre cooling facility for vegetables.
- Vibrators - 2 nos.

- Fluidized bed dryer for dehydrating. - 1 nos
- Hot water boiler - 1 nos
- Automatic form filling & sealing m/c - 3 nos
- Testing equipment

The total cost of machinery is estimated to be Rs.18.10 lakhs. The unit will also require miscellaneous assets such as furniture, fixtures, storage facilities etc. the total cost of these is estimated to be Rs. 1.25 lakhs. The total requirement of power shall be 50 HP

5. Raw Material and Packing Material

The basic raw material for the unit is different fresh vegetables. Depending on the availability of vegetables the product mix will change. Similarly the price would depend on the product mix. Thus without a firm mix it will be difficult to arrive at the actual price. On an average the raw material cost has been taken at Rs. 3000/- per tonne The unit will also require polythene bags for packing the finished product. The total cost of raw material and packing material at full capacity is estimated to be Rs. 14.50 lakhs. The total requirement is estimated to be 400 tonnes at 100% capacity. The price of raw material is taken at Rs. 3000 per tonne. At 60% capacity in 1st year the cost works out to Rs8.70 lakhs.

6. Land and Building

For smooth operation of the unit, it will require 500 sq. mts of open land and a built up area of 220 sq. mts. The total cost of land and building is estimated at Rs. 7.50 lakhs.

7. Manpower

For smooth functioning of the unit the requirement of man power is expected to be around 18 persons.

Sales person	**self**
Machine operators	4
Skilled Workers	4
Semi skilled workers	4
Helpers	6

The annual salary bill is estimated to be around Rs. 5.04 lakhs.:

8. Sales Revenue: (100% Capacity)

Selling price varies depending on the product mix quality and the availability of vegetables and demand. An average price of Rs 50,000/- per tonne has been taken the annual income at installed capacity is Rs 50.00 lakhs considering the yield of 25%.

9. Cost of Project:

Particulars	Rs. lakhs
Land & Building	7.50
Plant & Machinery	18.10
Other assets	1.25
Contingencies & pre-expenses.	5.05
Margin money	2.20
Total	34.10

10. Means of Finance

Promoters Contribution	10.10
Term Loan	24.00
Total	34.10

10. Profitability :(60% capacity)

Sales	30.00 (Rs. lakhs)
Raw material	8.70
Salary	5.04
Utilities	1.80
Stores & Spares	0.54
Repairs & Maintenance	0.60
Selling	2.25
Administrative expenses	0.66
Depreciation	3.35
Interest on T.L	2.64
Interest on W.C	0.50
Cost of production	26.08
Profit	3.92

12. Requirement of Working Capital

		Margin	W.C	Margin Money
Raw material	1 month	25%	0.40	0.10
Stock of finished goods	15 days	25%	1.40	0.35
Working expenses	1 month	100%	1.00	1.00
Sale on credit	1 month	25%	3.00	0.75
Margin money for W.C				2.20

13. Break Even Point 40%

14. Machinery Suppliers

1. M/S G.R.Engg works P. Ltd, Worli, Mumbai.
2. M/s Raylon Metal Works J.B.Nagar, Andheri(E) Mumbai.
3. M/S Laxicon Engg. Sita Bardi, Nagpur.
4. M/S Techno Equipment 31, Parekh Street, Girgaum, Mumbai.

G. Project on manufacturing of Pasta Products

Production Capacity	:	150 tpa

1. Product and its Applications

One of the popular pasta products is noodles made from tapioca flour and maida. These are thread like products, 0.22 to 0.40 mm in diameter. This product is becoming very popular due to increasing consumption of fast foods.

2. Market Potential

The demand for pasta food is over increasing due to growing trends towards fast foods particularly among younger generation, increase in the purchasing power of the people, convenience of preparation, scope for a number of recipes to suit individual's palate.

3. Basis and Presumption

1. The unit will work for 300 days per annum on single shift basis.
2. The unit can achieve its full capacity utilization during the 3rd year of operation.
3. The wages for skilled workers are taken as per prevailing rates in this type of industry.
4. Interest rate for total capital investment is calculated @ 12% per annum.
5. The entrepreneur is expected to raise 20-25% of the capital as margin money.
6. The unit would construct its own building.
7. Costs of machinery and equipment are based on average prices of machinery manufacturers.

4. Implementation Schedule

Project implementation will take a period of 8 months. Break-up of the activities and relative time for each activity is shown below:

Scheme preparation and approval	:	01 month
SSI provisional registration	:	1-2 months
Sanction of financial supports etc	:	2-5 months
Installation of machinery and power connection	:	6-8 months
Trial run and production :		01 month

5. Technical Aspects

5.1 Location

The plant can be located at any suitable place keeping in view the marketing convenience, availability of power, water and skilled manpower.

5.2 Process of manufacture

The noodle is manufactured in different sizes, hollow as well as solid and cooked in different methods. Some are made for cooking and others are for frying. The manufacturing method for frying quality noodles is as follows: The average moisture content of dry mixes is 10-11%. The three ingredients, viz. maida, starch and sodium bicarbonate are dry blended in a vertical mixer along with edible colour. Dough is made from the above blend by using boiled water wherein a part of the starch is gelatinized. The ingredients are mixed in dough mixer for about 12 to 15 minutes. The kneaded dough is transferred to a noodle making machine herein extruded material of desired shape and length is obtained by using an appropriate die and suitably adjusting the distance between the dye surface and cutting blade.

The moisture content of the product at this stage is about 33%. The cut noodles from the cutting machine fall on wooden trays. The product undergoes surface drying and becomes hard enough to be handled without sticking or being crushed. The moisture content of the pre-dried product at this stage is about 30%. The pre dried product is finally semi-dried. The moisture content of the product is around 17%. The product is exposed to steam for 15 minutes and subsequently dried to 10% moisture level.

6. Pollution Control

There is no major pollution problem associated with this industry except for disposal of waste which should be managed appropriately The entrepreneurs are advised to take "No Objection Certificate" from the State Pollution Control Board.

7. Energy Conservation

The fuel for the steam generation in the boiler is coal or LDO depending upon the type of boiler. Proper care should be taken while utilising the fuel for the steam production.

8. Production Capacity

Quantity	150 tpa
Installed capacity	700 kg/day
Optimum capacity utilization	70%
Working days	300/annum
Manpower	12
Utilities	
Motive Power	25 kW
Water	10 kL/day
Coal/LD oil	125 kg/day

9. Financial Aspects

9.1 Fixed Capital

9.1.1 Land & Building

Particulars	Amount (Rs. lakh)
Land 60 m 2	02.50
Built up area 150 m 2	11.00
Total cost of land and building	13.50

9.1.2 Machinery and Equipment

Description		Amount (Rs. lakh)
Vertical powder mixer cap 500 kg, Dough mixer, noodle making extruder, Wooden trays , boiler, weighing scales, buckets, handling equipment	:	05.50
Erection & electrification of machinery & equipment @10% cost	:	00.55
Office furniture & fixtures	:	01.75
Total	:	**07.80**

9.1.3 Pre-Operative Expenses Amount (Rs. lakh)

Consultancy fee, project report, deposits with electricity department etc	**02.00**

9.1.4 Total Fixed Capital Amount (Rs. lakh)

(9.1.1+9.1.2+9.1.3)	**23.30**

9.2 Recurring Expenses Per Annum

9.2.1 Personnel

Designation	No.	Salary Per month	Amount (Rs. lakh)
Factory Manager	1	12000	1.44
Supervisory staff	2	8000	1.92
Office Assistant	2	6000	1.44
Technician	2	5000	1.20
Skilled workers	1	3000	0.36
Unskilled workers	4	2500	1.20
			7.56
Perquisites @15 %			1.14
Total	**12**		**8.70**

9.2.2 Raw Material including Packaging Materials

Particulars	Qty (t)	Rate/t (Rs.)	Amount (Rs. lakh)
Maida	130	16000	20.80
Starch	20	14000	02.80
Salt, chemicals	LS	-	01.50
Packaging material	LS	-	05.20
Miscellaneous	LS	-	01.70
Total			**32.00**

9.2.3 Utilities

Particulars	Amount (Rs. lakh)
Power	0.90
Water	0.10
Total	**1.00**

9.2.4 Other Contingent Expenses

Particulars	Amount(Rs. lakh)
Repairs and maintenance @10%	0.61
Consumables & spares, others	0.55
Transport & travel	0.30
Publicity, postage, telephone	1.36
Insurance @1%	0.18
Total	**3.00**

9.2.5 Total Recurring Expenditure Amount (Rs. lakh)

(9.2.1 + 9.2.2 + 9.2.3 + 9.2.4)	**44.70**

9.3 Working Capital amount (Rs. lakh)

Recurring expenses for 3 months	**11.20**

9.4 Total Capital Investment Amount (Rs. lakh)

Fixed capital (Refer 9.1.4)	23.30
Working capital (Refer 9.3)	11.20
Total	**34.50**

10. Financial Analysis

10.1 Cost of Production (Per Annum) Amount (Rs. lakh)

Recurring expenses (Refer 9.2.5)	44.70
Depreciation on building @ 5%	00.55
Depreciation on machinery @10%	00.60
Depreciation on furniture @ 20%	00.35
Interest on Capital Investment @ 12%	04.15
Total	**50.35**

10.2 Sale Proceeds (Turnover) Per Year

Item	Qty (t)	Rate/t (Rs.)	Amount (Rs.lakh)
Noodles	150	45000	67.50

10.3 Net Profit Per Year

= Sales - Cost of production

= 67.50– 50.35

= Rs. 17.15 lakh

10.4 Net Profit Ratio

$$= \frac{\text{Net profit} \times 100}{\text{Sales}}$$

$$= \frac{17.15 \times 100}{67.50}$$

= 25.4 %

10.5 Rate of Return on Investment

$$= \frac{\text{Net profit} \times 100}{\text{Capital Investment}}$$

$$= \frac{17.15 \times 100}{34.50}$$

$$= 49.7\ \%$$

10.6 Annual Fixed Cost Amount (Rs. Lakh)

All depreciation	01.50
Interest	04.15
40% of salary, wages, utility, contingency	05.08
Insurance	00.18
Total	**10.91**

10.7 Break Even Point

$$= \frac{\text{Annual Fixed Cost} \times 100}{\text{Annual Fixed Cost} + \text{Profit}}$$

$$= \frac{10.91 \times 100}{10.91 + 17.15}$$

$$= \frac{1091}{28.06}$$

$$= 38.9\ \%$$

11. Addresses of Machinery and Equipment Suppliers

1. M/s Marvel Machines Pvt Ltd.
   140, Anna Salai, Saidpet
   Chennai - 600015
   Email: marvel12@vsnl.com
2. M/s Gee Gee Foods & Packaging Co. (P) Ltd.
   B -188/2, Savitri Nagar,
   Malviya Nagar
   New Delhi -110017
   Email: swantour@satyam.net.in

3. M/s Azad Engineering Company
   C-83, B.S. Road, Indl Area,
   Ghazibad -201009
   Email: azadeng@satyam.net.in/azadenggco@hotmail.com
4. M/s M. Son Industries
   D -33, Sector 2,
   Noida -201301
   Email: mson@nda.vsnl.net.in
5. M/s Sanjivan Industries
   A -13, Karimsheth Indl. Estate,
   Wagdevi Nagar, Near Valishali Nagar
   Dahisar (E), Mumbai -400068
   Email: sanjivanind@roltanet.com
6. M/s Pharamalab India Pvt. Ltd.
   Star Metal Compound,
   LBS Marg, Vikroli (W)
   Mumbai -400083
   Email: mkt@pharmalab.com
7. M/s. Mona Machinery Mfg. Co.
   Chandralok, 111 SD Road,
   Secunderabad (A.P.)
8. M/s. Khan Engg.
   Works, 5-5-274 Nampally,
   Near Gandhi Bhavan, Patel Nagar
   Hyderabad

12. Other Special Features

A careful selection of product mix is necessary based on the local market demand and availability of raw materials. The facilities can also be utilised to manufacture other pasta products with local recipes for fuller utilisation of capacity.

H. Project on Manufacturing of Fruit jam, Jelly and Marmalade

Quality and standards	:	As per FPO specifications
Production capacity	:	100tpa

1. Product and its application

Among processed fruits, jams, jellies and marmalades enjoy a predominant position. A large number of units are manufacturing these products to cater to the demand of domestic and export markets. The popular varieties are pineapple, mango, mixed fruit, guava, papaya, orange jellies and marmalades.

Jam is prepared from the fruit pulp by boiling with sufficient quantity of sugar to a moderately thick consistency, while the jelly is prepared from clear fruit extract. Marmalade is a fruit jelly wherein the slices of the fruit or peel shreds are suspended. The term is generally associated with products made from citrus fruits like oranges and lemons. The products are used as a bread spread and in bakery items. They can also be taken with chapati, dosa or similar breakfast foods to make them more appealing.

The packing ranges from 25 g, 500 g in bottles to 7 kg tins, depending upon the consumption need or bulk in 20 kg polyethylene-lined tins for commercial establishments. It is also packed in 15 g blister packs or 50 g, 100 g plastic cups.

2. Market Potential

Jams, jellies and marmalades share 17% of the total production of processed fruits and vegetable products. The demand is constantly increasing. The domestic market comprises defence sector, institutional sector, railways, airlines and regular channels of consumer stores and bakeries.

3. Basis and Presumptions

1. The unit will work for 200 days per annum on single shift basis.
2. The unit can achieve its full capacity utilization during the 3rd year of operation.
3. The wages for skilled workers are taken as per prevailing rates in this type of industry.
4. Interest rate for total capital investment is calculated @ 12% per annum.
5. The entrepreneur is expected to raise 20-25% of the capital as margin money.
6. The unit would construct its own building as per F.P.O. specifications.
7. Costs of machinery and equipment are based on average prices of machinery manufacturers.

4. Implementation Schedule

Project implementation will take a period of 8 months. Break-up of the activities and relative time for each activity is shown below:

Scheme preparation and approval: 01 month

1. SSI provisional registration : 1-2 months
2. Sanction of financial supports etc. : 2-5 months
3. Installation of machinery and power connection : 6-8 months
4 Trial run and production : 01 month

5. Technical Aspects

5.1 Location

The plant should be located in the vicinity of fruit growing/distribution centers keeping in view the marketing outlet. The other factors to be kept in view are ecology of the area and availability of transportation facilities (rail/road), cheap labour and other infrastructure facilities.

5.2 Availability of Raw Materials

Pineapple, passion fruit, grape, orange, ginger, lichi are available in large quantities. Firm, ripe and good quality fruits are to be used for the production. Preserved pulps can also be used. Other raw materials needed are sugar, pectin, citric acid, colour and flavours. Packing materials of different types are also readily available now-a-days. Innovative types of such materials are to be used to make the product more appealing.

5.3 Manufacturing Process

Good quality ripe fruits are selected and washed with water. They are peeled by SS knives and cut into small bits. The fruit bits are used directly or are mashed further and strained in a pulper. For jams, mashed fruit/pulp, for jelly, clear fruit aqueous extract and for marmalade, pulp and peel pieces (finely shredded) are to be used.

The cur fruit or pulp or fruit extract, as the case may be, it transferred to the steam jacketed kettle and heated to soften the fruit pieces. Sugar is added to this mass and heated further until it becomes thick in consistency. Colour, flavor and preservative are added at the end of the cooking process. Hot products are packed in bottles or plastic cups and cooled. The manufacturers have to take a license under FPO.

5.4 Quality Control and Standards: As per the PFA requirements

6. Pollution Control

There is no major pollution problem associated with this industry except for disposal of waste which should be managed appropriately. The entrepreneurs are advised to take "No Objection Certificate" from the State Pollution Control Board.

7. Energy Conservation

The fuel for the steam generation in the boiler is coal or LDO depending upon the type of boiler. Proper care should be taken while utilizing the fuel for the steam production. There should be no leakage of steam in the pipe lines and adequate insulation should be provided.

8. Production Capacity

Quantity	: 200MT
Optimum	: 700kg/day
Installed capacity	: 100%
Capacity utilization	: 70%
Working days	: 200/annual
Man power	: 18

9. Financial Aspects

9.1 Fixed capital

9.2 Land and Building (Amount Rs lakhs)

Land 500 sq.mtr	:	0.75
Built up area 150 sq. mtr.	:	4.50
——— Total cost of Land and Building	:	5.25

9.2 Machineries and Equipments

1.	(Pulper, SS kettle, fruit mill, bottle washing machine, bottle drier, plastic jar sealing machine, boiler, fruit washing tank, weighing scales, sauce pan, plastic cups, aluminium topped tables, SS knives, storage tanks)	: 6.00
2.	Erection and electrification @ 10% of machinery cost	: 0.60
	Office furniture & fixtures	: 0.80
	Total: ———	**7.40**

10. Pre-operative Expenses

Consultancy fee, project report, deposits with electricity dept	: 0.85
Total Fixed capital (1+2+3)	**13.50**

11. Recurring Expenses/ annum

11.1 Personnel

Designation	**No**	**Salary/month**	**(Rs.lakh)**
Factory Manager	1	10,000	1.20
Supervisor	2	6,000	1.44
Office Assistant	2	5,000	1.20
Technician	2	4,500	1.08
Skilled workers	3	3,000	1.08
Unskilled workers (10 months)	8	1,500	1.44
			7.44
Perquisites @ 15%Total :	18		1.12
			8.56

11.2 Raw Material Including Packaging Materials

Particulars	Qty.(MT)	RatePer mt.	Amount(Rs. lakh)
Fruit	80	06	4.80
Sugar	65	16	10.40
Citric Acid,Pectin	01	120	01.20
Colour, flavour	LS	LS	00.60
Jars/caps (500 g)	2 lakh	8	16.00
Total			**33.00**

11.3

Utilities	**Amount (Rs. lakh)**
Power	0.80
Water	0.01
Coal	0.19
Total	1.00

Steps for Commercializing Value-Added Food Business

Value-added food and agricultural products represent an obvious opportunity for agricultural producers to forward integrate from on-farm activities to serving the retail consumer. Likewise, entrepreneurs with a good product idea (e.g., a specialty salsa, grandmother's cake recipe, etc.) see commercialization of their idea as a means of becoming their own boss and owning a business. These are lofty and noble aspirations, and with proper planning, their dreams can come true.

A combination of general business rules and food-industry-specific rules must be followed to better ensure the successful development and long-term viability of a start-up business. While each segment of the food and agricultural products industry has specific rules, there are some general start-up guidelines that most small business professionals would recommend. The following represent a general set of steps to be taken in the commercialization of a value-added idea.

Trademark Protection

While this is probably not the very first step to be taken in starting a business, it is probably the most overlooked. Often those wishing to commercialize a food business concept have a specific name or logo in mind. However, too few check to see if some other business has claimed legal rights to a product name or logo. This oversight often results in lawsuits and/or the changing of product names/logos - a terrible thing for a business to do just as it is getting its product name recognized by consumers. Trademark protection is available at both state and federal levels. Check with state commerce authorities for pursuing protection of a name or logo within a state. However, if one wishes to market the product in more than one state, check with one of the USPTO's 80-plus patent and trademark depository libraries around the country. These libraries can provide free or at least inexpensive trademark searches, although one should probably use a patent/trademark attorney when filing for trademark protection.

Business Structure

A number of business structure possibilities exist: sole proprietorships, general and limited partnerships, limited liability companies, cooperatives and corporations are all common choices. The chosen structure will determine one's tax and legal liability status, as well as one's ability to raise capital. For example, a sole proprietorship is the easiest form of business to establish; in essence, the business is an extension of the owner. In a sole proprietorship, the owner pays taxes on business net proceeds in accordance with his/her tax bracket (and also self-employment taxes), but in the event of a lawsuit, the owner's personal assets are also at risk. On the other end of the spectrum, establishing a corporation requires the use of an attorney (and thus expense), but the resulting corporation is a separate legal entity and the owner's personal liability is limited to his/her investment in the business (assuming the business and personal bank accounts are maintained separately).

Facilities and Processing Equipment

This is an area where many start-up businesses run into problems. For almost any scenario, a health inspector will not approve of manufacturing products for retail sale in a person's home kitchen. Zoning rules, equipment requirements, utility requirements, licensing and permits are all issues related to mass production of a food product. Check with a local health inspector before buying, building or renting facilities for commercial food production. They can inform you of rules related to manufacturing facilities, necessary licenses and permits, equipment specifications and process requirements (i.e., rules to be followed for cooking, canning, preserving, etc.). Working with the health department inspectors from the very beginning will prevent a business owner from unwisely spending money at the starting stages and lessen the likelihood of fines or restrictions later. Additionally, if the costs of owning and operating a facility are too high for a start-up business, health inspectors and university food industry specialists may be able to help an entrepreneur find a co-packer, (i.e., a company that will, for a contracted price, manufacture and package the products for him/her).

Labeling Requirements

Nutritional labeling requirements, primary display panels, font requirements, declaration of contents, etc. are all issues related to labeling a food product. Small businesses are currently exempt from having to list the "Nutrition Facts" tables commonly seen on products in the supermarket if:

- Retail sales of the product are less than $50,000/year, or
- Retail sales of the business (i.e., for all products sold by the business) are less than $500,000/year.

However, it is commonly recommended that all businesses consider adding these tables since they would be required the moment the business exceeds the sales exemption level, and many consumers now look for that information on virtually all products they purchase. State and federal labeling requirements are generally similar, but one may want to contact his/her local health inspector for state labeling guidelines and for a simplified explanation of federal labeling rules. Also, universal product classification codes (commonly referred to as UPC codes or bar codes) are required for products sold in virtually every retail store and supermarket these days.

Marketing and Promotions

It is a common fallacy that a product is so good no promotion or marketing activities are necessary. In the competitive arena of food and agricultural products, these activities are essential. Many state agriculture or commerce departments/agencies operate programs to promote products manufactured within that state. Information is available on state programs that may be available to help promote a new food or agricultural product.These steps, along with common business activities such as developing financial projections, are all part of planning and commercializing a value-added concept. For more information, contact the agribusiness management and food industry specialists at the nearest land grant university.

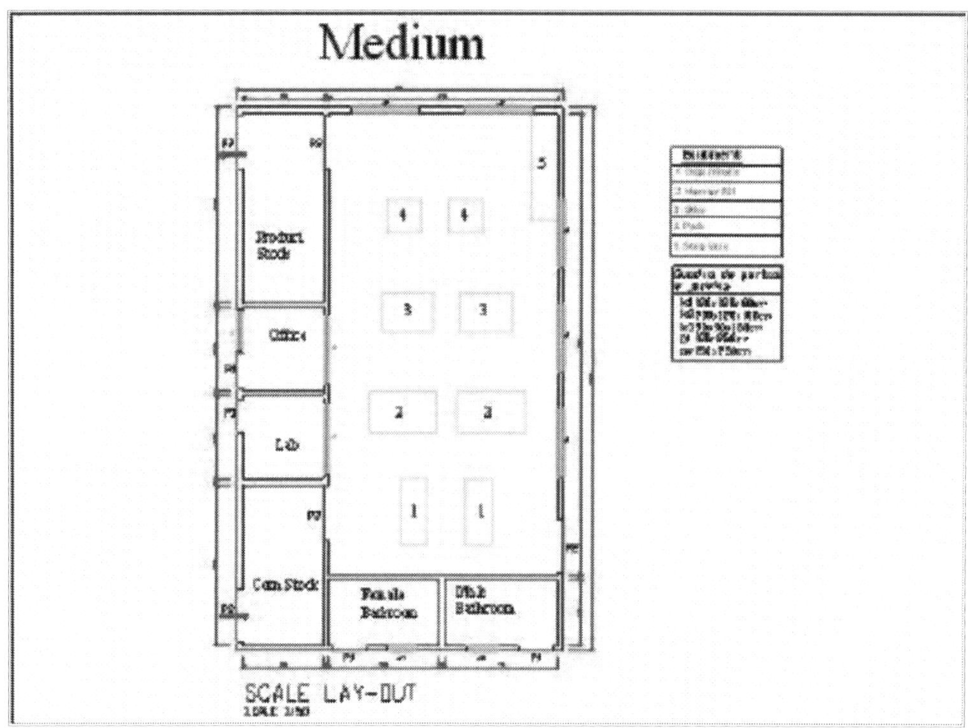

Figure 48: Layout plan for food processing unit

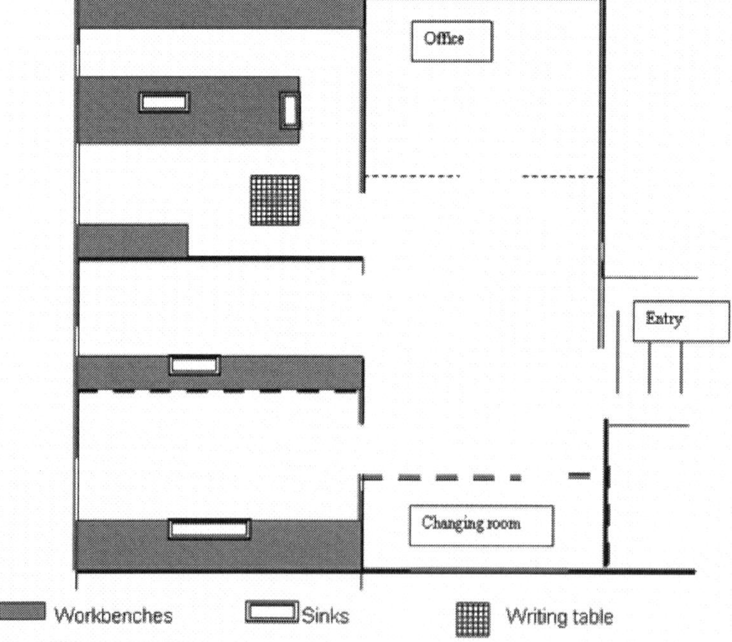

Figure 49: Food Analysis lab layout plan

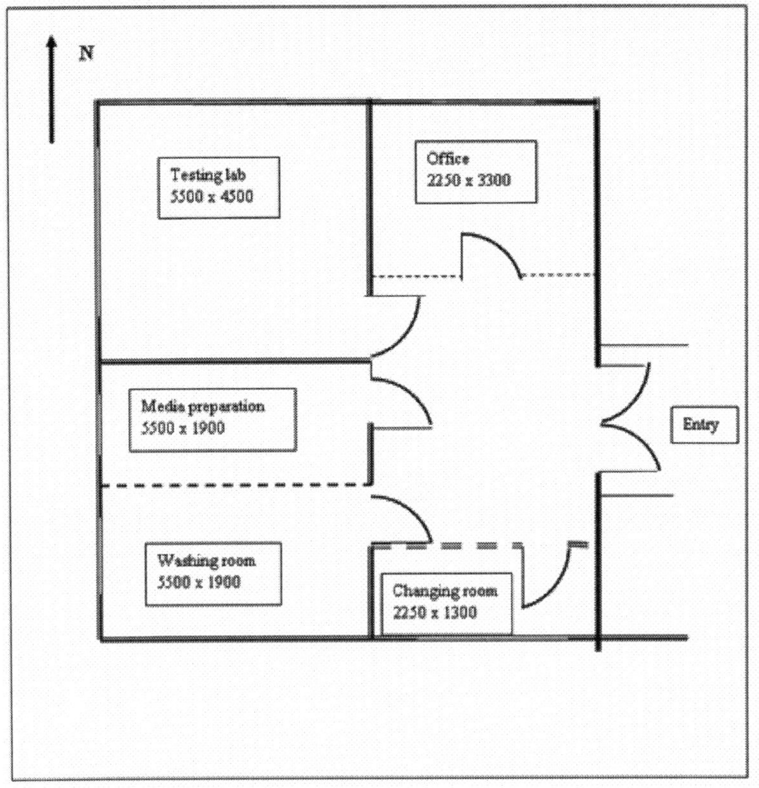

Figure 50: Food Microbiology lab layout plan

Table 44: Major thrust area for project preparation

- Asafoetida (Hing)
- Apple Juice Concentrate And Dehydrated Fruit & Vegetable
- Apple Fruit Juices with Canning & Bottling
- Atta Maida Suji & Wheat Bran (Roller Flour Mill)
- Areca nut (Betel Nut)Processing Unit
- Atta Chakki Plant
- Automatic Bread & Biscuit Plant
- Fresh Processed Frozen Vegetable Puree & Sauce
- Banana Plantation, Cultivation & Its By Products
- Baby Cereal Food
- Bread Plant (Semi- Automatic)
- Button Mushroom Cultivation, Processing & Canning
- Buffalo Meat Processing

- Baby Cereal Milk Powder
- Banana Powder
- Beer
- Biscuit Manufacture Buffalo Meat Processing
- Butter
- Baker
- Broiler Chicken
- Baking Powder
- Baby Cereal Food
- Betel Nut (Supari) processing
- Betel Nut Powder
- Blended Saccharin
- Bread Bread and Biscuits (Automatic Plant)
- Corn Oil (Maize Oil)
- Cottonseed Oil (Extraction & Refined)
- Cake Gel (Cake Improver)
- Cashew Apple Syrup Cum Orange/Lemon Squash
- Cashew Feni
- Cheese Analogues
- Chewing tobacco(Khaine) In Pouch Pack
- Khaini Chewing Tobacco (Raja Type)
- Chocolate
- Chocolate Drink
- Chocos (Ready To Eat Breakfast cereal food)
- Coconut And Cashew Feni
- Coconut Powder
- Coconut Squash Jam & Cream Custard Powder
- Candy Hard Boiled
- Carbonated Beverage
- Cashew Feni
- Coconut Powder
- Cocoa Butter
- Cocoa Butter from Cocoa Mass
- Chewing Gum
- Chilli Oil
- Chillies (Processing & Grinding)

- Chilli Sauce
- Condensed Milk (sweetened)
- Cones For Softy lce Cream Corn Flakes
- Custard Powder
- Cube Sugar
- Curry Paste
- Canning of Rasgullas In Metal Cans (Indian Sweet)
- Confectionery Industry
- Chewing Gum & Bubble Gum
- Chocobar
- Cold Drinks (Soft Drinks)
- Condensed Milk (Sweetened)
- Carbonated Beverage
- Dehydration And Pickling Of Oyster Paddy Straw Mushroom
- Drum Stick Powder
- Dall Mill
- Dehydrated Vegetables French Fries & Allied Potato Products
- Fruit Juice (Mango) In Tetrapack
- Fruit Juice In Plastic Cups
- Fruit Juice In Tetrapack
- Fruit Juice, Jam And Jellies
- Fruit Pulp & Juices
- Fruits Concentrates (Rasanatype)
- Fruit Juice Making And Packing In Plastic Container (Pouches)
- Fruit Processing (Jam & Jellies)
- Fruit Juice Powder
- Fruit Cakes
- Flour Mill (Disc Machine)
- Fish Canning
- Fish Farming
- Food Colour (Coal Tar Based)
- Food Dehydration
- Fruit Juices Making & Packing in Plastic Containers (Pouches) Fruit Juices, Mango, Pineapple, Apple, Lichhi
- Frozen Foods
- Frozen Finger Chips

- Fast Food Parlour
- Ginger Cultivation & Storage
- Ginger Paste In Pouch, Black Container
- Grape Wine
- Gur From Cane (Export Quality)
- Garlic Powder
- Grinding of Dry Red Chillies Grape Wine
- Glucose-D Powder
- Gram Dal & Flour Mill with Modern Automatic Plant
- Green Peas Processing And Canning
- Gram Dal
- Grinding of Rock Salts & Iodization
- Guar Gum Powder
- Hard Sugar Candy
- Health Drink (Cocoa Beverages in Granules Form)
- Hing (Asafaeoefida)
- Honey Roasted Peanut
- Hard Boiled Candy
- Honey Processing & Packing
- Instant Coffee
- Instant Noodles
- Instant Tea
- Iodised Salt (Free Flowing) From sea Water
- Ice Making Plant
- Ice Cream Stabilizers
- Idli mix, Dosa mix, Sambhar mix, Vada mix, Gulabjamun mix, Tomato Soup mix (Instant food)
- Instant Coffee
- Indian Made Foreign Liquor
- Instant Jellies with Different Flavours
- Instant Ice Cream Mix in Various Flavours
- Instant China Grass
- Ice Cream & Ice Candy
- Pickles (Various types) Instant Tea
- Insoluble Saccharin
- Iodized Salt
- Iodized Salt from Crude Salt

- Instant Coffee & Instant Tea (Premixed with Sugar & Milk)
- Jack Fruit Processing
- Jam Group, Tomato Sauce and Tomato Ketchup
- Khaine (Chewing Tobacco)
- Kimam
- Khandsari Sugar
- Katha & Cutch
- Lecithin From Sunflower Oil
- Liquid glucose From Maize And Maize Oil
- Lecithin (Soya Based)
- Locally Made Foreign Liquor
- Liquid Glucose And Its By products
- Liquid Glucose From Maize
- Maize & Its Products
- Maize Cultivation & Its By-Products
- Food Colour, GMS, Caramel, Reactive Dyes & Pigments
- Micro Nutrients Mixture Solid
- Mushroom Processing & Canning
- Macaroni Manufacturing
- Macaroni, Sapghetti, Vermicelli & Noodles
- Mango Fruit Bar
- Mango Juice
- Mango Papad (Aam Papad)
- Mango Pulp
- Roasted Salted Cashew kernel From Cashew Nut
- Milk Powder, Pasteurised Milk, Butter, Cheese & Ghee
- Milk Product Cheese
- Macaroni
- Maida Mill
- Maize Starch
- Making & Canning of Rasgullas in Metal Cans
- Mango Beverages
- Margarine Fat
- Masala
- Mayur Brand Type
- Chewing Tobacco

- Mini Sugar Plant
- Meat Extraction
- Milk Powder
- Milk Preservation and Marketing to Whole Sales (in pouch packing) by UHT Technique
- Milk Toffee
- Milk Soluble & Insoluble Powder
- Modern Rice Mill
- Mango Seed Powder
- Mango Processing & Aam Papad
- Mineral Water
- Mineral Water, Soda water & Pet Bottle
- Milk Powder
- Mineral Mixture
- Malting Plant
- Mutton Processing
- Natural Coloured Oil (Turmeric Colour & Oil)
- Non-Dairy Whipping Cream
- Non Dairy Whipping Cream
- Nicotine from Tobacco Waste
- Orange Crush
- Pasteurised Milk
- Piggery/Meat/Chicken Processing
- Pork Products
- Papad & Bari
- Parboiled Rice Mill
- Pasteurised Milk & Cheese
- Potato Chips
- Peanut Butter
- Pickles Murabbas (Veg. & Non Vegetarian)
- Pineapple Juice Manufacturing & Canning
- Poha (Chiwra)
- Potato And Onion Flakes
- Potato Chips (Different Type Recipe & Flavoured)
- Potato Granules
- Potato Powder (Automatic)
- Potato Starch

- Processed Food & Spices
- Processing And Packaging Of Snack Food
- Processing Of Fruits & Vegetable
- Palm Oil Crushing Unit
- Pan Masala
- Paneer (Cheese)
- Paneer From Milk (Soya Milk)
- Pasteurization of Milk
- Peanuts-Roasted
- Peanut Butter
- Pickles and Murabbas (Vegetarian and non Vegetarian Pickles)
- Piggery Products (Meat)
- Piggery/Meat/Chicken Processing
- Pineapple Flavour For Bakery And Soda Water
- Preservation of Raw Mango Juice
- Potato and Onion Flakes
- Potato Chips
- Pop-Corn
- Potato Starch
- Potato & Onion Powder
- Potable Alcohol from Grains & Damaged Fruit
- Processed Food
- Processing & Retail Packing of Food Grains, Pulses, Split Pulses, Spices, Pickles
- Pan Masala, Meetha, Saada, Zarda Making(Gutkha) and Packing
- Purification of Casein
- Refined oil (Cottonseed, Ground Nut Oil & Sunflower Oil)
- Refining Of Edible Oil
- Rice Flake (Poha)
- Rice Flake (Poha)
- Refining of Mustard Oil
- Refining of Salt & Manufacture of Table Salt
- Rice and Corn Flakes
- Rice Flakes
- Roller Flour Mill (50 MT/Day)
- Spices (100% E.O.U)
- Soyabean Oil From Soya Bean Seed & Cattle Feed

- Starch From Tapioca
- Sacarine (Soluble & Insoluble)
- Sattu Manufacturing Unit
- Slaughter House Beef Processing
- Soft Drink Concentrate
- Soyabean Meat
- Spices (EOU)
- Spices (Grinding Base Spices With Packaging In Bags)
- Starch & Allied Products From maize
- Sugar Candy (Mishri)
- Sugar Pellets
- Sugar Plant (Crushing 5000 TPD)
- Soft Drink (Cola, Orange, Lemon, mango, Ginger, Clear Lemon)
- Sweet Aroma Of Betal Nut
- Salt Licks For Cattle
- Salted Biscuits
- Salted Groundnut
- Scented Supari
- Softy Ice Cream Cone (fully automatic imported plant)
- Softy
- Snuff Factory
- Soft Drinks
- Soft Drinks (Non Carbonated) Mango, Litchi, Guava, Pineapple Flavours, Frooti Type Tetrapacked
- Soft Drink (Non Carbonated) in Tetra pack Frooti Type
- Soft Drink Concentrate & Essence (Orange, Cola, Lime Pineapple) Flavour For Bakery And soda water
- Soyabean Bariyan (Nugget Nutrella Type)
- Soyabean Products
- Soya Milk And Paneer
- Soda Water & Sweet Drinks
- Spices
- Spices (Kitchen King, Degi Mirch, Chat Masala, Raita Masala)
- Squashes From Pineapple, Orange,
- Lemon Starch From Tapioca
- Sterilisation of Double Toned Milk

- Sugar Plant
- Sugarcane Juice Preservation
- Sugar Cubes
- Suji
- Sugar Candy
- Sweet Meals
- Sweet Scented Supari
- Tamarind Juice Concentrate
- Tea Plantation , Cultivation & Processing
- Tea Processing
- Tabacoo Flavouring compound
- Tamarind Juice Powder
- Tamarind Pulp From Tamarind
- Tea Packaging Industry
- Tomato Paste (Tomato Concentrate)
- Turmeric Powder
- Tamarind Juice Concentrate
- Tamarind Juice Powder
- Tea Industry
- Toffee Candy & Milk Chocolate
- Toffee & Sweets (Automatic Plant)
- Toffee & Candy Making (Semi-Automatic Plant)
- Tomato Powder
- Tomato Products
- Turmeric Processing
- Ice Cream
- Upgradation Of Salt
- Vanaspathi Ghee (Hydrogenated Vegetable Oil)
- Vermicelli
- Vinegar & Malt Vinegar For Industrial & Domestic Use
- Vodka From Potatoes
- Wheat Puff
- Washing of Ginger
- Whisky
- Wine Distillation From Fruit And Cashew Nuts (Cashew Feni)
- Wheat Puff

- Yogurt In Plastic Caps
- Yeast From Molasses
- Tobacco Zarda Zafrani Baba type
- Caffeine from Tea Waste
- Moistureless, Free Flow Iodized Salt
- Pectin from Raw Papaya
- Fiber From Banana Plant & MGF. Of Bags Like Jute Bags
- Honey Processing & Packaging
- Mini Flour Mill
- Roller Flour Mill
- Refined oil (Cottonseed, Ground Nut Oil & Sunflower Oil)
- Refined Oils (Sun Flower Oil, Ground Nut Oil and Cottonseed Oil)
- Rice Bran Oil
- Capsicum Oleoresin
- Solvent Extraction Plant
- Dextrose Powder From Potato
- Dairy Farm And Dairy Products
- Dehydrated Raw Mango
- Dehydrated Garlic Flakes and Granulated Powder
- Vanilla Plantation, Cultivation & Processing

References

Allison, Penelope. Pompeian Households: An Analysis of the Material Culture, University of California Los Angeles, 2004

Cool, Hillary. Eating and Drinking in Roman Britain. Cambridge University Press, 2006.

Curtis, Robert. Ancient Food Technology. Leiden; Brill, 2001. "Umbricius Scaurus of Pompeii", Studia Pompeiana and Classica in honor of Wilhelmina Jashemski, vol 1: Pompeiana. Ed. Robert I. Curtis: New Rochelle, NY, Caratzas, 1988.

Grainger, Sally. Cooking Apicus: Roman Recipes for Today. UK Prospect Books, 2006.

http//www.emeraldinsight.com/searc?

http//www.dabur.com/product-foods-fruit%20read

http//www.pepsico.com/Brands/Tropicana-Brand.html.

http//www.pepsiindia.co.in/Brand/Beverage/Tropicana.ospx

Kotler and Philip, " Marketing Management prentice, Hill of India, edition-IX

Ramaswami, V.S and Namakumari,S " Marketing Management" MacMillan, edition-II

Mission Statement, Yale Sustainable Food Project (http://www.yale.edu/sustainablefood/about_mission.html)

Sambasiva rao and Umesh mishra, 2009, Agri. Business project-An option for diversification Indian journal of Fertilizer

Zeitfracht Medien GmbH
Ferdinand-Jühlke-Straße 7
99095 Erfurt, Deutschland
produktsicherheit@kolibri360.de